I0100686

PRACTICALS IN BASIC ENTOMOLOGY

PRACTICALS IN BASIC ENTOMOLOGY

Dr. T.V. Sathe
Professor in Entomology,
Deptt. of Zoology,
Shivaji University, Kolhapur

Dr. P.M. Bhoje
Sr. Lecturer, Zoology Deptt.
Y.C. College, Warna Nagar, Kolhapur

Miss Vaishali S. Kolekar
Deptt. of Zoology,
Shivaji University, Kolhapur

2014

Daya Publishing House®
A Division of
Astral International Pvt. Ltd.
New Delhi – 110 002

First Published: 2005
Reprinted: 2014

ISBN 9789351243786 (International Edition)

Publisher's note:

Every possible effort has been made to ensure that the information contained in this book is accurate at the time of going to press, and the publisher and author cannot accept responsibility for any errors or omissions, however caused. No responsibility for loss or damage occasioned to any person acting, or refraining from action, as a result of the material in this publication can be accepted by the editor, the publisher or the author. The Publisher is not associated with any product or vendor mentioned in the book. The contents of this work are intended to further general scientific research, understanding and discussion only. Readers should consult with a specialist where appropriate.

Every effort has been made to trace the owners of copyright material used in this book, if any. The author and the publisher will be grateful for any omission brought to their notice for acknowledgement in the future editions of the book.

Published by : **Daya Publishing House®**
A Division of
Astral International Pvt. Ltd.
– ISO 9001:2008 Certified Company –
4760-61/23, Ansari Road, Darya Ganj
New Delhi - 110 002
E-mail: info@astralint.com
Website: www.astralint.com

Laser Typesetting : **Vaishnav Graphics & Systems**
Delhi - 110 055

Digitiality Printed at : **Replika Press Pvt. Ltd**

Dedicated to

Asawari T. Sathe

(M.B.B.S. III)

CONTENTS

1.

INSECT COLLECTION AND PRESERVATION

Broadly, insects are grouped into three categories:

1. Useful insects 2. Harmful insects and 3. Neutral insects.

Honey bees, silkworms, lac insect, parasitoids, predators are manageable as natural resource. Hence, they are useful. Some insects cause damage to various kinds of plants, stored poducts, stored grains, live stock and man. Hence, they are destructive and harmful insects.

Neutral insects are those insects whose role, either beneficial or harmful is not known to science. A very large number of insects are neutral or unidentified. The useful and harmful insects may have their combine share about 5 per cent in total insect world. Up-to-date only 35 per cent insects have been identified, thus, entomologist have big task before them to identify and describe the remaining insects. Insect biodiversity has attracted the attention at global scenario. Because, biodiversity can help sustainable development of the region or a country. Therefore, insect collection, identification and their correct preservation has great importance.

Objectives of the Collection and Preservation of Insects

1. To keep the record of insects from different habitats and localities.
2. To study the insect species with respect to morphology, taxonomy, evolution, biology, behaviour and molecular biology, etc.
3. To find out the systematic position of insects in class insecta.

Material Required for Collection and Preservation of Insects

1. Sweeping net
2. Insect or butterfly net
3. Pond net
4. Specimen tubes
5. Hand lens
6. Entomological forcep
7. Insect store box
8. Insect rearing cage
9. Light traps
10. Water trap
11. Suction trap
12. Insect killing bottle
13. Insect setting board
14. Entomological pins
15. A note book.

1. Sweeping net

It comprises 40 cm long handle of wood or metal with screw arrangement to metal ring. It has 65 cm long bag made up of baft with 30 cm diameter mouth. It is useful for collecting most of the insects including grasshoppers, bugs, moths, beetles, etc.

2. Insect or butterfly net (Fig. 1)

A insect net is designed with handle 24" long attached to a iron ring of about 18" diameter. A collecting bag of 30" in depth made up of ordinary mosquito netting cloth is attached to the iron ring. This is easy for collecting most of the large and medium sized insects including butterflies, moths, ichneumon flies, wasps, bees, etc.

3. Pond net

It is made up of nylon netting but, it has longer handle than sweeping net. It is more or less similar to insect net. By dipping and pulling the net quickly in water and to water surface may catch

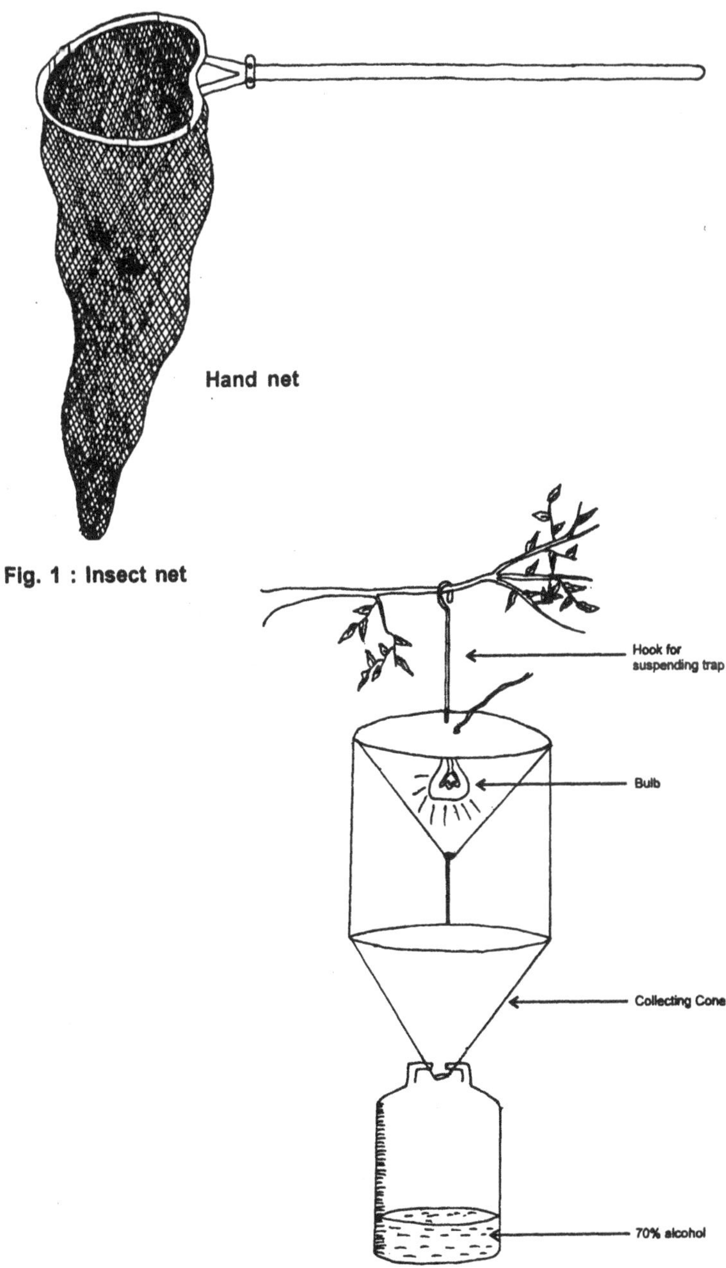

Fig. 1 : Insect net

Fig. 2 : A Light trap

several aquatic insects. Thus, pond net is useful to collect aquatc insects.

4. Light traps (Fig. 2)

Standard light traps are available in the market, those are supplied by scientific suppliers.

In general, light trap comprises a light source like bright electric bulb, 200 watts or tube light providing white background or mercury vapour lamp. A typical light trap is shown in Fig-2 which is useful for collecting night insects (nocturnal). The insects get attracted to light source and falldown in collecting cone while flying around the light source. Then the insect desend down into collecting jar containing 70 per cent alcohol. It is better to remove all the insects from collecting jar within 24 hrs but may be kept for a week or more in survey study.

5. Water trap

It is shallow tray of size 12.5 × 20 cm deep containing water and a drop of "teepol". Inner surface of the trap should be bright or white for attracting the insects.

6. Suction trap

It is typical devise of collecting insects which is useful for sucking the small insects such as mosquitoes, small moths like *Corcyra cephalonica* and other microinsects.

7. Sticky traps (Fig. 3 & 4)

It comprises a cylinder covered with sticky material like grease, oil, osliko, etc. The trap may be arranged at suitable height. Sticky trap may be quadrangular sticky panel of white or yellow colouration (Fig. 4). It may be fitted in the field at various heights for collecting insects which are active at both, day and night.

8. Specimen tubes

Glass specimen tubes of various sizes are useful for storing small insects in 70 per cent alcohol or 4 per cent formalin. They are also useful for handling the small insects in the laboratory.

9. Entomological forcep (Fig. 5)

This is special type of forcep. It has typical bend shape. Hence, useful in pinning process of insects.

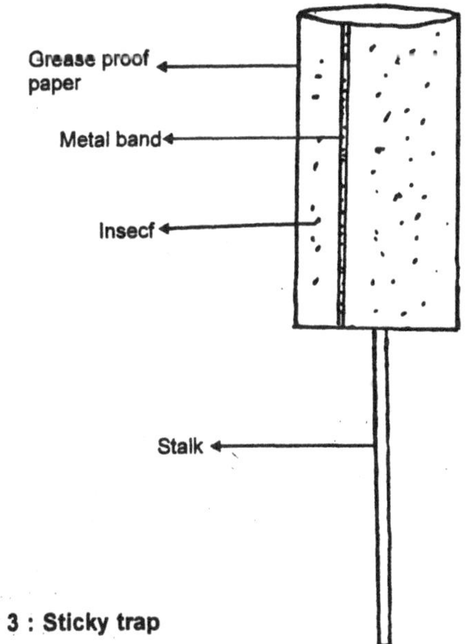

Fig. 3 : Sticky trap

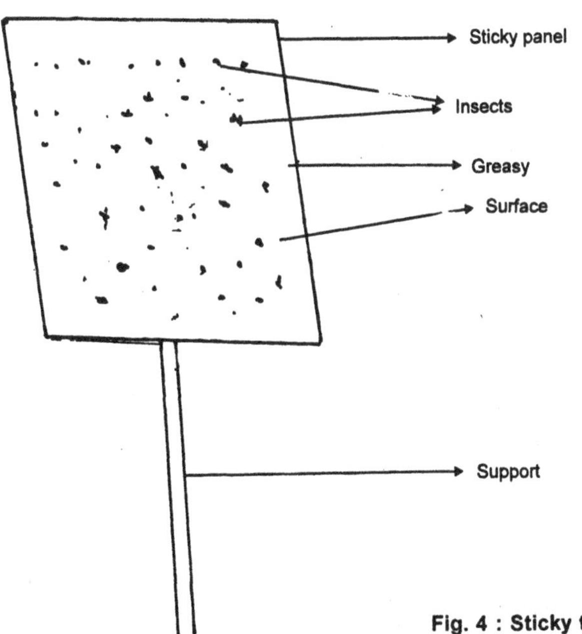

Fig. 4 : Sticky trap

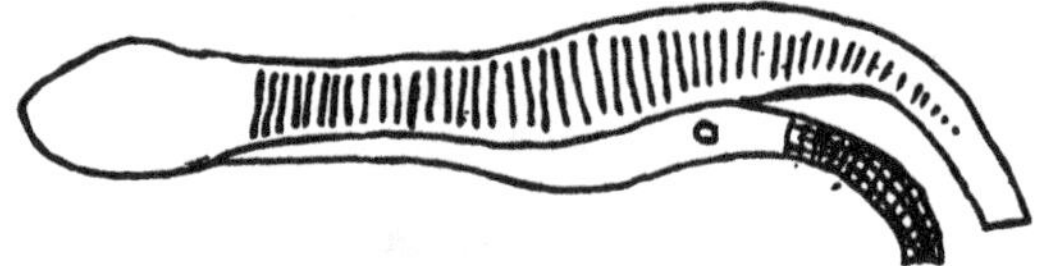

Forcep (Entomological)

Fig. 5 : Entomological forcep

10. Hand lens

Hand lens is essential in field collection. It is useful for magnification of insects and thus, identification of insects.

11. Insect store box (Fig. 6)

Correctly pinned insects are to be stored in insect store box for preserving them safely for longer period. Insect store box is made up of seasum wood such that their joints are insect and dust proof. The bottom as well as top portion is covered with cork sheet for pinning purpose. The insects are mounted on the basal portion. It is strictly advised that only the best specimen are kept in the insect store boxes. Incorrectly pinned, rusty and damaged insects should not be kept in an insect box.

Fig. 6 : Insect store box

12. Insect setting board (Fig. 7)

Insect setting board is made up of three wooden stripes and screws for adjusting space between two stripes as per the requirement of the size of insect. It is useful for setting the insects in proper position. The setting board is shown in Fig. 7.

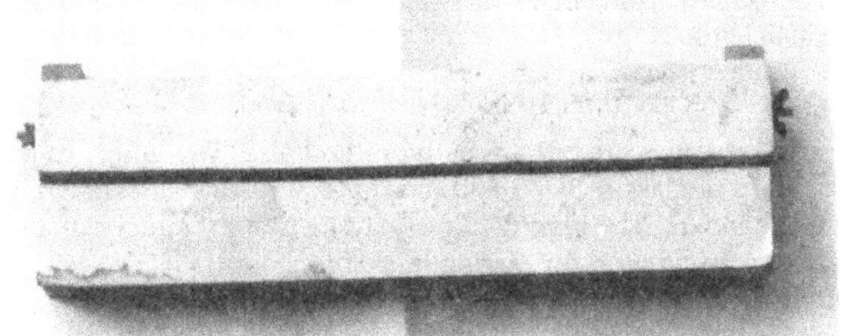

Fig. 7 : Insect setting board

13. Entomological pins (Fig. 8)

Entomological pins are longer than ordinary pins and also smaller in diameter than ordinay pins. They are silver coated. Therefore, helpful for avoiding rusting to pinned insects. Entomological and ordinary pins are shown in Fig. 8. Entomological pins are rust proof and made up of mixture of brass and nickel.

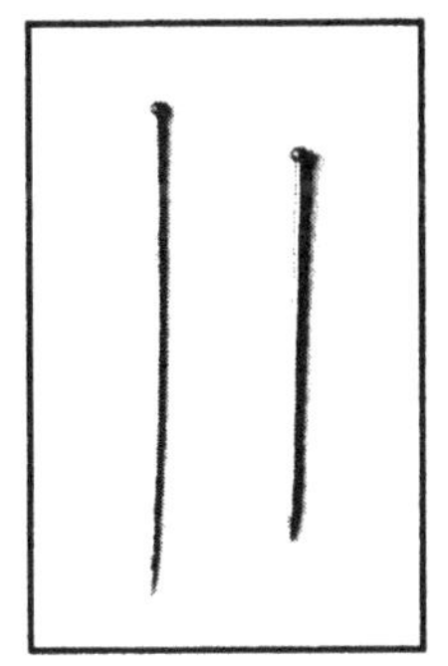

Fig. 8 : Entomological pin (left)

14. Killing bottle (Fig. 9)

Method for Preparation of Killing Bottle

Take an ordinary wide mouth bottle having ½ kg capacity with tight iron screw cap. A mixture of KCN or NaCN and plaster of paris 1 : 3 is then filled in bottle to ⅓ of its capacity. Then arrange the pieces of blotting paper on the surface of mixture so that the moisture present in the mixture/bottle will be absorbed. After two days bottle will be ready for use.

Precautions Regarding Killing Bottle

It should not be kept open which provides poisonous gas (hydrocynic gas). Bottle must be labelled as "POISON".

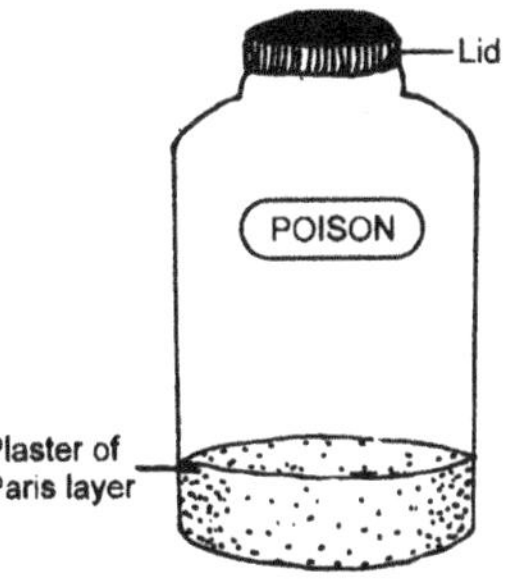

Fig. 9 : Poison bottle

Broken pieces of the bottle must be disposed by buring deep in soil. It is advised that lepidopterous insects should not be mixed with others while keeping in killing bottle. Do not keep insects for longer time in bottle.

15. Insect cages (Fig. 10 & 11)

Insect cages are of various types and size. Two insect cages made up of glass & wood and mosquito net sleeve are shown in Fig. 10 and 11. The rearing cages may be of iron wire mesh and wooden frame used for rearing butterflies and moths or keeping adults of large insects in experimental study.

Fig. 10 & 11 : Insect rearing cages.

Collection of Insects

Insects is diverse group of animal kingdom. Hence various types of collection methods are adopted. Best time for collection is evening and morning because, most of the insects they become active during this period probably for their basic needs *i.e.,* Food, Shelter and Mate. Some of the collection methods of insects are given in the text.

1. Hand picking

Hand picking is best and simple method for collecting insects such as grasshoppers, bugs, caterpillars, beetles, etc. They may be picked up by hand from plants or other objects and placed in plastic container/killing bottle.

2. *Neting*

With the help of butterfly net, moths, butterflies, and many other flying insects are collected. While, sweepnet is specially useful to collect insects such as grass-hoppers, plant bugs etc. Aquatic insects such as ranatra, scorpion bug, and others are collected with the help of pond net.

3. *Trapping*

Light traps, sticky traps, water traps, suction traps, pitfall traps, pheromone traps, etc are used for trapping the insects and their periodic collections in survey studies.

4. Shaking the flowers and other plant parts

Thrips are always associated with flowers. They are collected by shaking flowers on plain paper. Later, thrips are picked to specimen tube by camel hair brush dipped into 70 per cent alcohol. Likely aphids, jassids, mealy bugs, are collected by shaking small twigs of the plants. A beating tray is kept under branch and the branch is beated. Insects falling in the tray are collected and preserved/used for further study. The beating trays are marketed by scientific equipment suppliers.

Killing of Insects

For killing insects, first see the index of endanger insects, rare insects and banned insects and avoid those from the collection and preservation point of view. However, common insects are killed by killing bottle. As highly poisonous chemicals are banned, therefore, for killing insects, use cotton bolls soaked into chloroform/ethylacetate/tetrachlorethane. The poisoned cotton bolls are kept at the base of killing bottle. However, insects should not be kept in the bottle for moe than 20 minutes. Mount and set the insects within 24 hours.

Preservation of Insects

1. *Pinning*

Pinning is best method for preserving insects. This method is useful in handling insects in taxonomical studies. Pinned insects are stored quite safely than other insects. Entomological pins are of various sizes. As a rule ⅔ potion of pin must be below the pinned insect and ⅓ above.

How and Where to Pin the Insects

Orthopterans should be pinned through the mesothorax, back parts of thorax, slightly right side of middle line.

Hemipterans are pinned through mesoscutellum slightly to right side of middle line (Fig. 12).

Hymenopterans are pinned through the centre of mesothorax (Fig. 13).

Coleopterans are pinned through the base of right forewing (elytra) (Fig. 14).

Lepidopterans are pinned through the middle of thorax (mesothorax) (Fig. 15).

Dipterans (Fig. 16) are pinned through mesothorax from ventral side so that the pin should not emerge from dorsal side as chaetotaxy should not damage on dorsal surface which is useful criteria for identification.

Small insects are pinned by micropins (small sized pins).

Spreading and Positioning of Insects

With the help of spreading board/setting board insects such as butterflies, moths, dragonflies, tripsipterans, etc are properly spread on the board. Antennae should direct forward, abdominal appendages ovipositor, etc should direct backward. Wings of lepidopterans should be spread properly. Hind edge of forewing should be at right angle to the body and hing wing is appropriately matched with the forewing. Wing setting is done by pinning paper stripes on the spreading board. Your teacher will demonstrate the setting of insect on setting board.

Carding

The small insects are placed on a white rectangular card 5×8 mm or 5×12 mm. The size varies with the species. Below this card data labelled card is also provided which contain the information about the collection i.e. Identification of insect, host plant, date of collection, name of collector, etc., (Fig. 17). Small triangular cards are also used for pinning the small insects. Its detail is shown in Fig. 18.

Mounting of Insects

Collected insects if not mounted immediately, becomes hard

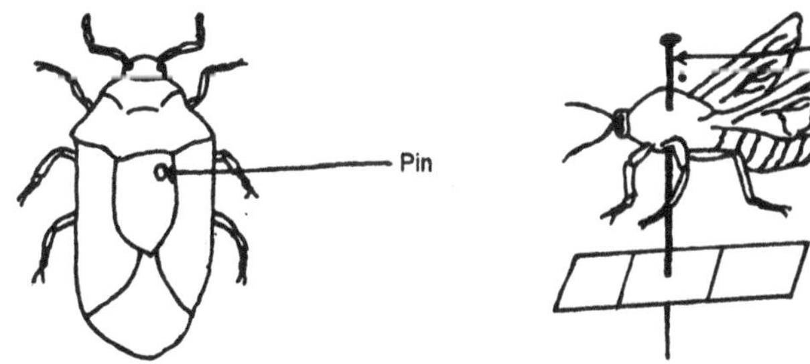

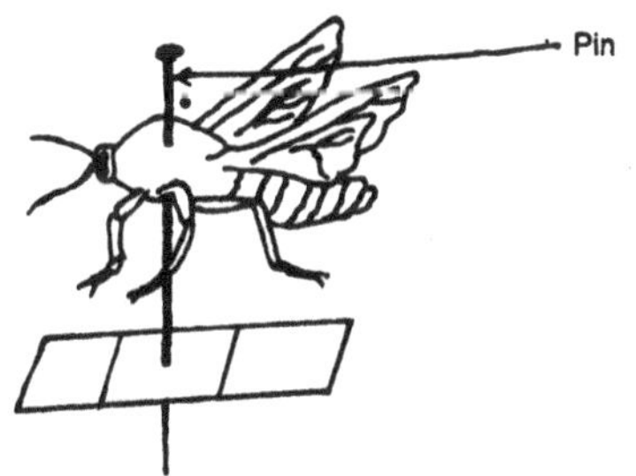

Fig. 12 : Hemipteran bug

Fig. 13 : Hymenopteran wasp

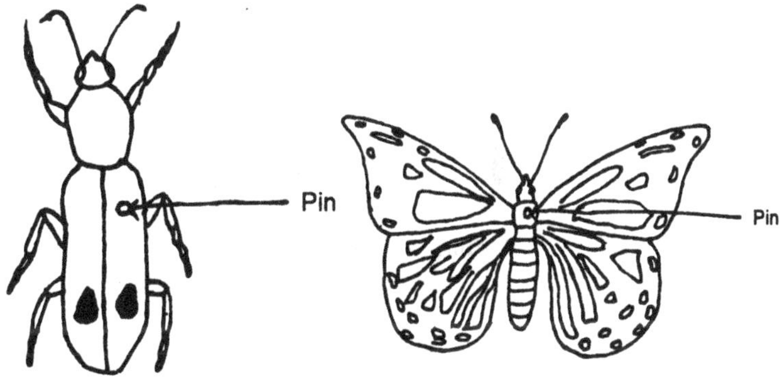

Fig. 14 : Coleopteran beetle

Fig. 15 : Butterfly

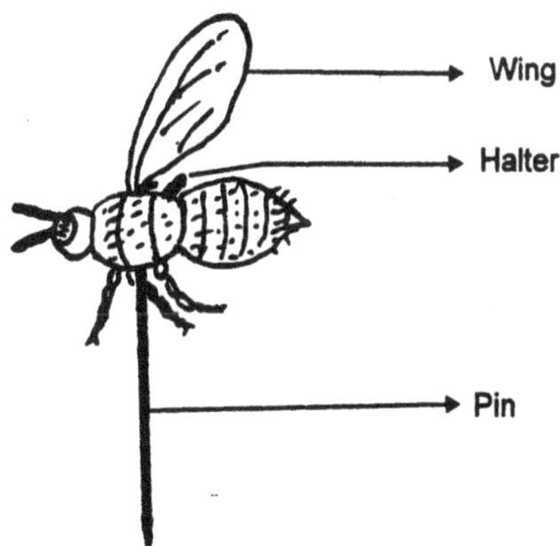

Fig. 16 : Dipteran fly

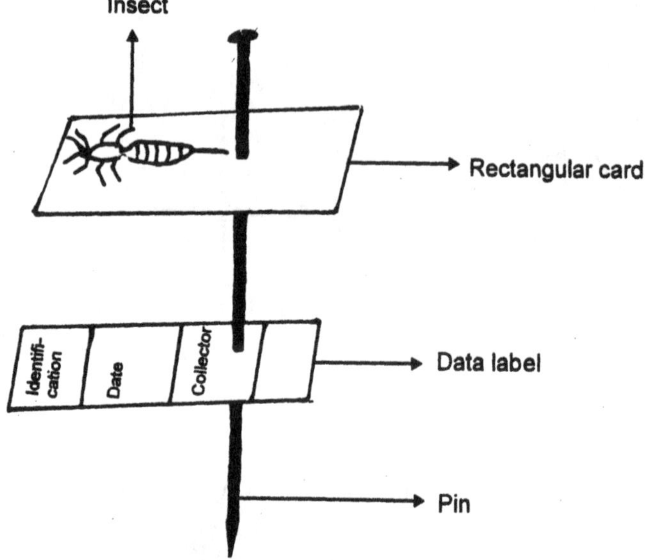

Fig. 17 : Carding with rectangular card

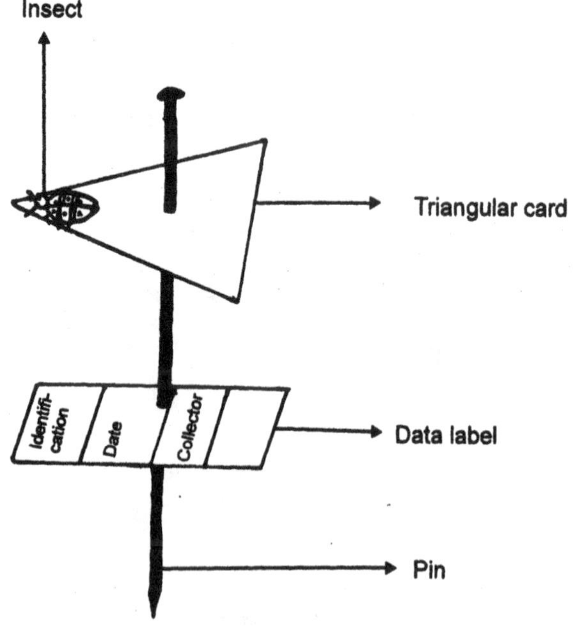

Fig. 18 : Carding with triangular card

and such insects are relaxed before mounting. Relaxation is done in a wide mouth airtight jar in which with moist sand covered blotting paper is spread. In this jar add few drops of carbolic acid for preventing mould formation. The insects can be temporarily stored in such jar for 1 to 2 days.

Drying of Insects

After proper spreading and pinning, the insects are dried at oven or drying chamber at 60° C. Drying will avoid future fungal or animal attack. After drying insects are transferred to insect box. Napthelin balls wrapped in a piece of paper should be kept in corner of box for avoiding fungal growth on preserved specimen.

Microinsects and Their Preservation

Microinsects are collected with the help of camel hair brush and kept in specimen tubes containing 70 per cent alcohol. After passing through the various alcoholic grades, lower to higher (dehydration), insects are cleared by xylene and mounted on slide with the drop of DPX or Canadabalsm. Give the proper position to the insect on slide. Place the insect laterally (Fig. 19) arrange anteunae in forwarding direction, forwarding, wings must forward

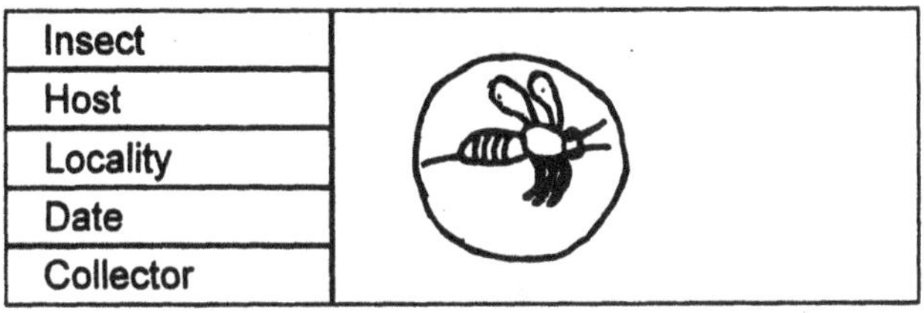

Fig. 19 : Insect preservation on slide

dorsally, legs ventrally and ovipositor, etc. backwardly. Flat insects are placed on slide dorsoventrally. Spread the wings laterally, anteunae and forlegs forwardly and mid and hing legs backwardly. Label the slide properly by identification, host record, locality, date of collection and name of collector.

2.
Detection of Chitin in Insect Cuticle

Insect required

Living cockroach (*Periplanata americana*) (Blattaria : Blattaridae)

Apparatus

500 ml Beaker, petridish, glass rod, cavity block, watch-glass, thermometer – 200° C, glycerine bath, spirit lamp, test tubes, etc.

Chemicals

Glycerine, dist. water, alcohol grades 90 – 30%, saturated KOH solution, 5% CH_3COOH, 2% Iodine, 1% H_2SO_4 .

Importance and Theory

Chitin is nitrogenous polysaccharide polymer and major part of the insect body wall (cuticle). According to Wigglesworth (1968) chitin is protein complex of cuticle with empirical formula $(C_6H_{13}O_5H)x$ and high molecular wt. The chitin molecules are composed of N-acetyl glycosamine and β-glucosine linkage. Single chitin molecule contains hundreds of N-acetylamine units. "Kitobiose" is a single unit of two adjacent of n-acetyl glucose amines. The units can join together in a linear fashion forming branched long chain. Endocuticle contain main constituent of chitin and vary with the species in per cent.

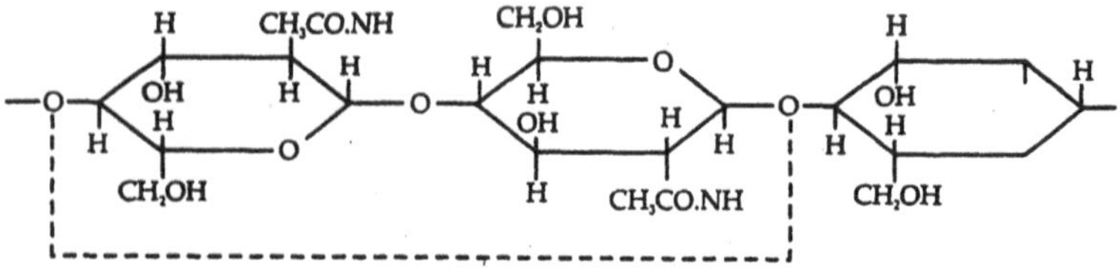

Properties

Chitin is soluble in dilute acids and concentrated alkalies. It is also soluble in water, ethers, alcohol, all solvents including acetic acid and H_2SO_4. It forms iodate and chitoson sulphate when treated with H_2SO_4. Chitin is soluble in conc. mineral acids and it is hydrolyzed to lower saccharides when reacted with alkali at high temperature.

Procedure

The procedure includes following steps :

1. Take out a living cockroach and remove few pieces of tergites or sternites from it, clean the pieces.
2. Remove all other parts such as tracheae, muscles, etc. from these pieces so as to clean the matter.
3. The pieces are then taken into test tube containing saturated KOH solution.
4. Uniform and indirect heating of pieces on glycerine bath containing thermometer 200° C.
5. The test tubes containing pieces and KOH solution then heated continuously till the temperature rises to 140°C.
6. At 140°C the colour of pieces becomes faint slowly.
7. Heat the test tubes continuously up to 155°C, at this temperature cuticle becomes decolourised completely.
8. Continue the heating of pieces till the temperature goes to 160°C.
9. KOH solution starts boiling at 160°C. Heating is stopped when material becomes decolourised completely. Arrange temperature 160°C constant for continuous heating and this process is repeated for about 15 minutes.
10. In decolourised material chitoson is left. Chitoson is colourless and transparent.
11. After proper heating the test tube is taken out from glycerine bath and cooled at room temperature.
12. Take out chitoson (Pieces) in clean watchglass. Some pieces are washed in cavity block with degraded alcohol, from absolute to 30 per cent for 2 – 5 minutes each.

13. Take out chitoson pieces on clean slide/watchglass and treat with 1% H_2SO_4. The material will change to brown.
14. The same matter is treated with 0.2% iodine solution for observing violet colour. The violet-colour to the matter will indicate the presence of chitin.

Confirmatory Test of Chitin

Treat the chitoson pieces on slide with 3% acetic acid, the matter will be dissolved immediately adding 1%. H_2SO_4 to this matter will develop white ppt. This will confirm the presence of chitin.

Special Precautions

1. Heating the material beyond 160°C should be avoided, residue may not be left in the test tube if heated beyond 160°C.
2. For hoping positive results, acetyl bond, must be broken completely by heating.
3. Direct heating is avoided, heating should be uniform and indirect.
4. Adopt gradual cooling process.

Results

The intense violet colouration confirm the presence of chitin in the matter.

3.

Detection of Waxy Component in Insect Cuticle

Experimental Insect

Living black ants (*Formica* sp.) (Order : Hymenoptera).

Apparatus

Compound microscope, cavityblocks (2), Camel hair brush, etc.

Chemicals

Liquid paraffin, absolute alcohol (methylated spirit), xylene, etc.

Importance and Theory

A large waxy layer (Fig. 20) is present in the epicuticle of the insects. The waxy component may have different characteristics in different insects. It is soft greasy in cockroaches and waxy, pale yellow, noncrystalline white in the larvae of sawfly *Athalia proxima*, *Nematus* sp. and *Pieris brassicae* (Cabbage caterpillar) while, in stored grain pests like *Tenebrio* sp. it is hard and crystalline.

If the insect is treated with xylene, the waxy component of cuticulin is removed and the waxy layer presents a flow of water from the insect body. Water bubbles are different from air bubbles by the characteristics that the air bubbles explode immediately when they come to the air surface. The water has great affinity towards alcohol therefore, it comes out on large scale and attracted toward the alcohol layer. Wax is a mixture of paraffins of the probable order C_{26}-C_{31} and esters of n-alcohols and acids. The waxy layer may range in thickness from 0.1 μ to 0.4 μ.

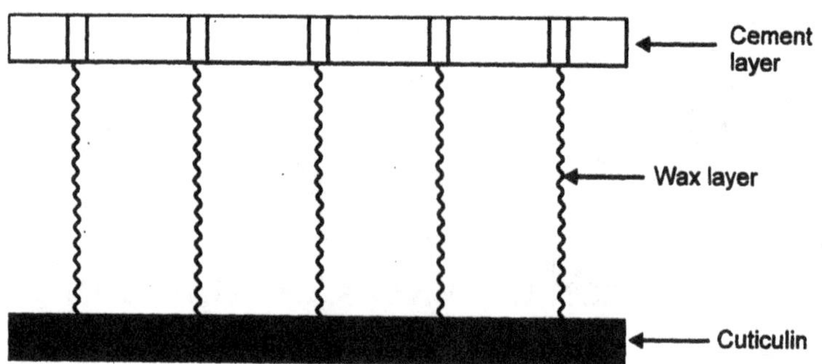

Fig. 20 : Waxy layer in insect cuticle

Procedure

Take a cavityblock (I) containing few drops of xylene. Then submerge a live black ant *Formica* sp. in it for 15 minutes. The xylene will react on the waxy component of ant cuticle/epicuticle. Then arrange one more cavityblock (II) containing a mixture of methylated spirit and liquid paraffin. These two chemicals doesn't mix with each other. Now submerge xylene treated ant into this cavityblock (II) containing mixture of LP and MS/absolute alcohol. After few minutes observe the cavityblock under microscope. It will be observed that–

1. Bubbles emerge out of the cuticle.
2. Number of bubbles increase with respect to increase in time.
3. By rising temperature small bubbles becomes large bubbles.
4. These bubbles are of alcohol water.

Results

The water bubbles coming from cuticle proves presence of waxy layer in insect cuticle which dissolves on treating with xylene.

4.
DETECTION OF URIC ACID IN INSECTS

Experimental Insect

Blister beetle (*Zonabria pustulata*) (Coleoptera : Meloidae).

Source

Malpighian tubules and Rectum.

Requirements

A. *Apparatus*:Dissecting box, mortar and piston, filter papers, spirit lamp, test tubes, slides, coverslips, etc.

B. *Chemicals*:Uric acid, $AgNO_3$ solution, conc. HNO_3, dil. NH_4OH, Na_2CO_3, NaOH solution, distilled water etc.

Importance and Theory

Uric acid is the most important nitrogenous component of the urine. It contains less hydrogen than other nitrogenous compounds excreted and thus useful in water conservation. As a free acid or as Ammonium salt it is highly insoluble but, eliminated with a very little water. It helps refilling the water supply. It helps the insect in egg and and pupal development where there is no source for refilling the water supply. Nitrogen is excreted as uric acid with 85.8% in silkworm. In dried excreta of *Tineola* it is 28%' while in *Tenebrio* it is 50% and in *Antheraea perneyi* 26.2%. It is in solution when sufficient water is available but in scarce, it is in a crystalline spheres. The crystalline sphere may have the range of diameter from 1μ to 60μ or more. The uric acid also have other nitrogenous constituents.

Procedure – I

1. Dissect out Blister beetle and remove the malpighian tubules present at junction of hindgut and midgut of the alimentary canal.
2. The malpighian tubules are cleared by removing tracheae, fat bodies and other elements.
3. The malpighian tubules are then taken in a cavityblock containing little quantity of distilled water.
4. Prepare a homogenous paste of M.T.
5. Add to this a little quantity of $Na_2 CO_3$ solution and stirr a little while.
6. Take few drops of $AgNO_3$ on the filter paper (Fig. 21) then add a drop of homogenized extract on it.
7. Yellowish brownish or black coloured ppt is developed (Fig. 21).
8. This indicates the presence of uric acid in the extract.

As due to the reduction of $AgNO_3$ ppt is developed.

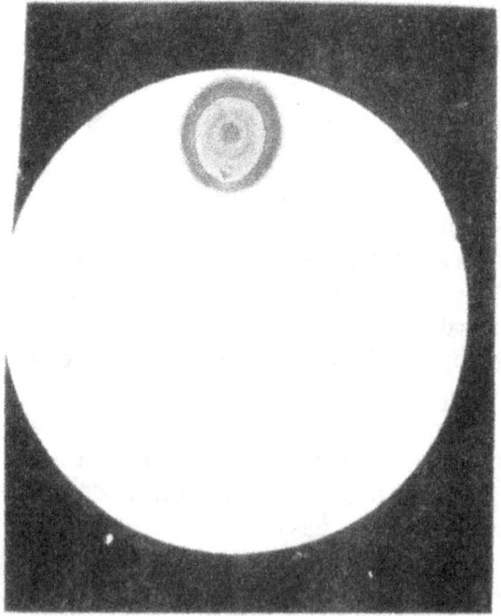

Fig. 21 : Filter paper with black-brownish PPT

Result

Yellowish or brownish colouration confirms the presence of uric acid in the extract.

Procedure – II

1. Dissect out live Blister beetle.
2. Take out rectal content from rectum of blister beetle and place on clean slide.
3. Prepare a smear of the extract on slide.
4. Dry the slide.
5. A few drops of conc. HNO_3 are added and evaporated it to dry.
6. Heat the slide.
7. Content will turn brown due to the formation of alloxanthine.
8. Add a drop of NH_4OH – the reduced turns purple in colour.
9. Dry the slide with NH_4OH.
10. Examine the slide under microscope.
11. Blue spots confirms the presence of uric acid in the extract.

Reaction

$$\text{Uric acid} + \text{Conc. } HNO_3 \xrightarrow{\Delta} \text{Alloxanthine (Purple)}$$

$$\text{Alloxanthine} + NH_4OH \longrightarrow \text{Murexide (Blue).}$$

Result

Uric acid is present in rectum of the insect. (Blister beetle).

5.

To Measure the Rate of Spiracular Movement in a Given Insect

Experimental Insect

Cockroach, *Periplanata americana* (Blattaria : Blattidae).

Requirements

Dissecting microscope, stop watch, dissecting tray, pins, dissecting box, etc.

Importance and Theory

Spiracles are respiratory organs in the insects. Spiracles are outlets or openings of the respiratory system. During respiration spiracle maintain a typical rythm of opening and closing. Excess of CO_2 and O_2 deficiency acts as a stimulant for control of spiracular movement. Spiracular opening is induced by oxygen deficiency and carbonic acid accumulation controlled by oxygen deficiency alone results in rapid fluttering in spiracles. CO_2 in air dysponia shows 120 frequency per minute. CO_2 in O_2 dysponia attains the frequency of 45 per minute. Respiration rates vary with the species and situations. At 15% CO_2, the primary movement increase to 90 – 120 per minute while, at 20 – 30%, respiration rate becomes 150 – 180 per minute and the spiracles no longer close.

Procedure

Take a dissecting microscope and mount the experimental insect under microscope so as to locate thoracic spiracle(s) very clearly in the focus. Give the proper position to insect for noting observations on the rythm of opening and closing of spiracles under room conditions. After positioning insect, start the stopwatch and take the observations on the opening and closing rythm of spiracles. Count the opening and closing rate of spiracles per minute. The observations are repeated atleast ten times and mean value is taken as the rate of spiracular movement.

Observations*

Sr. No.	Minutes	No. of opening and closing
1	1	30
2	2	45
3	3	40
4	4	37
5	5	57
6	6	50
7	7	47
8	8	67
9	9	70
10	10	40
Average		48.3

Result

Rate of spiracular movement in a given insect is 48.3/min.

* Hypothetical.

6.
Detection of Loss of Moisture During Respiration

Experimental Insects

Cockroach *Periplanata americana* (Blattaria : Blattidae)

Requirements

Apparatus: Hygrometer, chamber or plastic container of 2 litre volume, stopwatch, etc.

Importance and Theory

Spiracles are involved in respiration and protection of insect from loss of water since it is provided with sphincters. The rate of loss of water or moisture vary with the species and thus opening and closing rate of spiracles. In rat flea *Xenopsylla cheopes,* the rate of loss of water doubles when the spiracles kept open by exposure to 5% carbon dioxide. While, in a mosquito *Anopheles* sp the loss increased to 23.2 per cent when spiracles were kept open to 2 – 3% CO_2. In several beetles water is lost much more rapidly when subelytral space was kept open but not so quickly when the subelytral space was kept humid. Thus, spiracles are involved in loss of moisture.

Procedure

Take a clean plastic container of volume 2 litres. Then insert hygrometer in it and note the reading. Name this reading as "R_1". Later, take known number (10) of cockroaches in the container and note the reading on hygrometer. A reading is taken at every five minutes. The readings are terminated when there is no increase in reading in hygrometer for long period. Then stop the experiment and a final reading is taken giving the name as "R_2" and also note the time. Then calculate the amount of moisture given out by each insect for every time minutes. A graph is plotted against increase in hygrometer reading and time.

Observation Table

Sr. No.	Time in minutes	Moisture in %
1	5	10
2	10	17
3	15	20
4	20	22
5	25	24
6	30	24
7	35	24
8	40	24
9	45	24
10		

Observations :

1. Moisture in chamber before experiment R_1 = 6%
2. Volume of chamber 2 liters
3. No. of insects-10
4. Moisture of chamber after experiment R_2 = 24%
5. Wt. of H_2O in 1 lit. of air before experiment = .0003 gm/lit.
6. Wt. of H_2O in 2 lit of air before experiment 0.006gm/2 lit.
7. Wt. of H_2O in 1 lit of air after experiment 0.012 gm/lit.
8. Wt. of H_2O in 2 lit of air after experiment 0.024 gm/2 lit.
9. Wt. of H_2O expelled by insects = 0.000144 gm/2 lit.
10. Wt. of H_2O expelled by insects at 9 min. = 7.75 gm/ 2 lit/min.
11. Wt of H_2O expeled by insect in 1 min. = 0.75 gm/ 2 lit./min.
12. Wt of water expelled by each insect in 1 min = 0.075 gm/ 2 lit. min.

Calculations

1. Moisture in chamber (Volume of chamber 2 litre) before experiments, $R_1 = 6\%$
2. Number of insects 10
3. Moisture of chamber after experiment $R_2 = 24\%$ in 2 lit.
4. Weight of H_2O in one litre of air before experiment $= \frac{R_1}{1000} = \frac{3}{1000} = 0.003$gm/lit.
5. Weight of H_2O in two litre of air before experiment (x) = A × 2 × gm/2 lit. = 0.003 × 2 = 0.006 gm/2 lit.
6. Weight of H_2O in one litre of air after experiment $= \frac{R_2}{1000} = \frac{12}{1000} = 0.012$ gm/lit.
7. Weight of H_2O in 2 litre of air after experiment = (y) 0.012 × 2 = 0.024 gm/2 lit.
8. Weight of H_2O expelled by insects (y × x) = 0.024 × 0.006 = 0.000144 gm/2 lit.
9. Weight of H_2O expelled by insects at 9 minutes = 0.75 gm/2lit/min.
10. Weight of H_2O expelled by insect in one minute = 0.075 gm/2 lit/minute

∴ Weight of H_2O expelled by each insect in one minute $= \frac{0.75}{10}$ mg/2 lit/minute = 0.075 gm/2lit. min.

7.
Demonstration of Uptake of Dye by Malpighian Tubules in a Given Insect

Experimental Insects

Cockroach	*Periplanata americana* (Blattaria : Blattidae).
Grasshopper	*Heiroglyphus banian* (Orthoptera : Acrididae) or
Blister beetle	*Zonabria pustulata* (Coleoptera : Meloidae).

Requirements

Apparatus

Shallow dish beewax tray, dissection box, dissecting binocular/microscope with good lighting, etc.

Chemicals

Indigo-carmine/0.01% Neutral red.

M.T. Number and Species Diversity

Grasshoppers, and nymphal cockroaches contain 60 or more malpighian tubules, Lepidopterous caterpillars contain 6 tubules while, Dipterous larvae (*Drosophila*) contain only 4 tubules, the pairs unite together and open into the rectum by single pore in Drosophila.

Importance and Theory

Malpighian tubules are elongated, tubular bodies which extends from the junction of mid and hind gut and are found in insects. These vary greatly in number, from a pair to over 100 in different species and also vary with the forms. Nitrogenous waste, uric acid, are passed into the tubules in solution. Excess water is resorbed in the tubules or in the rectum, making the urine and fecal material in a more or less dry form. Thus, malpighian tubules have excretory function.

There are several types of dyes which enter into the lumen of malpighian tubules. In cockroach indigocarmine is excreted by only some part of malpighian tubules. In certain insects like *Rhodnius* the above dye is excreted only by the upper segment. Similarly, neutral red enters the lumen by the same route. However, later, it is taken up by the lower segment, deposited and stored in the cells. Many dye enters into the lumen through cells. Thus, sometimes, they store some part of dye in their bodies. The stored dyes may be later, absorbed by the lumen in many cases.

Procedure

Select any of the above given insect for experiment. Anaesthelize the insect with CO_2 or ether or chloroform before dissection. Pin the insect in a shallow dish beewax tray and dissect away the dorsal or ventral abdomenal wall and carefully lift or tease away structures obscuring the M.T. Tease out a malpighian tubule into a few drops of saline. Add a small amount of neutral red solution and observe the uptake of dye by the tubule cells. Also observe its appearance in the M.T. lumen. An evidence of red dye will be observed by the presence of crystals in M.T.

Results

Crystals are seen in malpighian tubules.

8.

Histological Preparation of Haemocytes from the Haemolymph of a Given Insect

Experimental Insect

Cockroach, *Periplanata americana.* (Blattaria : Blattidae)

Requirements

Apparatus

Slides, coverslips, microscope, water bath, beaker 500 ml, thermometer, lamp etc.

Chemicals

Geimsa's stain or Leishman's stain, 2% ringer solution, 2% Versene, alcohol grades, xylene, DPX, etc.

Importance and Theory

Haemocytes are an important constituent of the blood of insect and are loose cells present in the blood in body cavity of insect. These cells are generally located on the surface of various organs and also found circulating freely in the haemolymph or blood. Haemocytes are present in lumen of heart certain insects such as *Periplanata americana, Pieris brassicae, Apis indica,* etc but not in certain insects such as Bed bugs, etc. In most insect species, the haemocytes are found either independently in blood or in aggregations (Sathe and Rokade, 1994). They change their shape as per the need (Wago, 1982). The number of circulating cells vary with the species from 15,000 to 2,75,000 per cubic millimetre, a cockroach averaged 30,000 per cubic millimetre.

Sites of haemocytes: Tissues, surface of various organs, blood, along out surface of heart, wing vesels, wing veins, heart lumen, etc. In certain cases haemocytes are sedentory, eg. *Corethra* larva.

The haemocytes perform the functions such as the ingestion of small solid particles (phagocytosis), resistance to micro organisms, blood coagulation, connective tissue formation, resistance to metazoan parasites, haemocytes and intermediary metabolism, immunity, giant cell formation, etc.

Procedure

A cockroach is immersed for about 2 – 3 minutes in hot water bath (temp. 53 – 54°C but, not more than 60°C) for removing haemocytes from their aggregation sides. Cut the tip of antenna and press the cockroach gently for a drop of haemolymph from the cut point. A monolayer prepared degreased slide should be kept ready with a drop of 2% versene ringer solution as haemocyte fixative. Then take a drop of haemolymph in fixative. The slide is then air dried and then stained with either Leishman's stain or Giemsa's stain for 10 minutes. Then remove the slide and wash it with running tap water and allow to dry at room temperature. Make the dehydration of the matter by brief bath of acetone. The stained slides are cleared in xylene and then finally mounted in DPX.

Steps Involved in Technique

1. Dip the insect in hot water (temp. 53°C – 54°C) for abou' 2–3 minutes.
2. Take out blood.
3. Smear on slide.
4. Add dilution fixative (2% versene).
5. Dry it with lamp (bulb).
6. Stain it.
7. Dehydration.
8. Prepare slide.
9. Mount under microscope.
10. Observe the characters of haemocytes in slide.
11. Identify the haemocytes.

1. *Haemolymph dilution Fluid*

2% versene-ringer solution

+ traces of methelene blue.

2. *Adherent medium*

One part egg albumin

+4 parts 2% versene-ringer solution.

3. *Chemical fixative*

Disodium EDTA (Versene) should be used as chemical fixative. It is prepared by dissolving 2 gm of disodium EDTA crystalline powder in 100 ml solution of 2% versene. Ringer solution. (insect ringer).

4. *Preparation of insect ringer's solution*

Dissolve following in distilled water to make the volume 1,000 ml (P^H – 7 to 7.2)

Nacl – 9.8 g

Kcl – 0.77 g

$Cacl_2$ – 0.5 g

Na_2HCO_3 – 0.18 g

NaH_2PO_4 – 0.01 g

Dextrose – 1.0 g

5. *Giemasa's stain*

Giemsa powder – 0.5 g

Methyl alcohol – 50 mls

Glycerine – 50 mls

Acetone – 10 mls.

Working solution – (0.1 ml in 0.9% distilled water).

6. *Leishman's stain*

Leishman's powder – 0.5 gm

Methyl alcohol – 50 mls

Glycerine – 50 mls

Acetone – 10 mls

9.
Qualitative Count of Haemocytes in a Given Insect

Experimental Insects

Cockroach, *Periplanata americana* (Blattaria : Blattidae) or Gram pod borer, *Helicoverpa armigera* (Lepidoptera : Noctuidae)

Requirements

Apparatus

Microscope, slides, coverslips, needle, waterbath, beaker 500 ml, thermometer, etc.

Chemicals

Leishman's stain, Giemsa's stain, xylene, DPX, alcohol grades, etc.

Procedure

For the fixation of haemocytes, the insects are emmersed foı 1 – 2 minutes in a hot water bath of temperature 55 to 60°C. A grease free slide is kept ready for taking haemolymph drop emerging from tip of antenna. With the help of scissor a cut is taken to the terminal portion of antenna. Immediately haemolymph will start emerging as a drop from cut point of antenna. It is taken on slide and a thin/monolayer is prepared and then dipped into either Giemsa's stain or Leishman's stain for 10 minutes. Then the slide is washed with tap water and dehydrated. After giving a wash of xylene, the slide is mounted with DPX and observed under microscope. Following haemocytes are observed under microscope.

For *H. armigera* blood is taken from cutting prolegs of the larvae.

1. Proleucocytes (Fig. 22)

These cells are small rounded or oval in shape with deeply stained cytoplasm. These are abundant in blood smear specially in ribonucleic acid and nucleus. A huge bulk of basophilic granules are found in cytoplasm. They are abundant in young Lepidopterus larvae like Gram pod borer. They are found undergoing mitosis in all insects and have refered as young growing individuals.

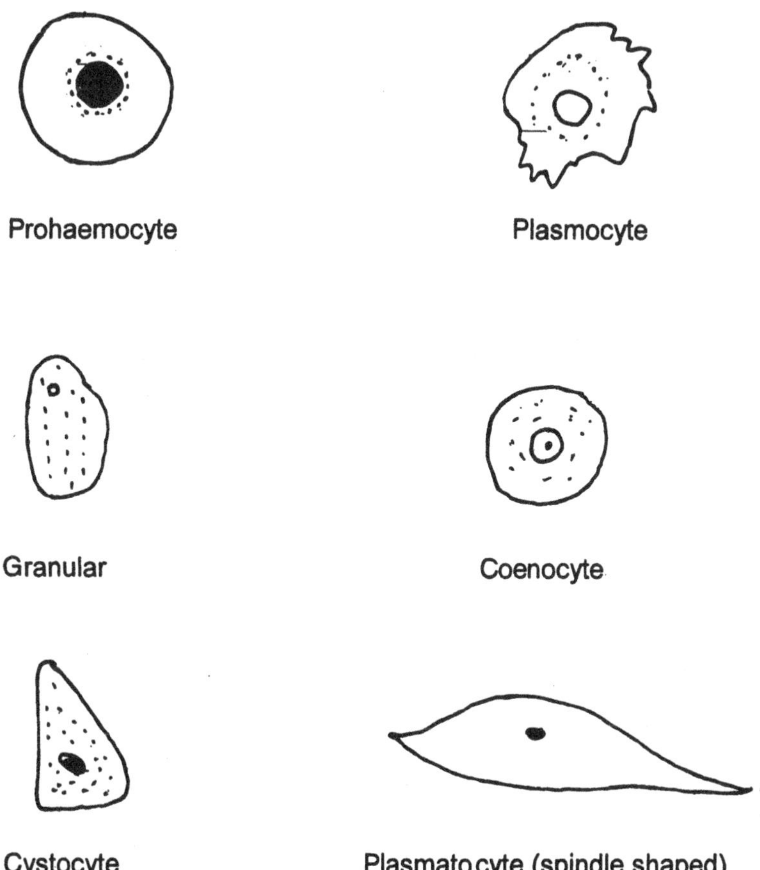

Fig. 22 : Types of haemocytes in cockroach

2. *Plasmatocyes (Fig. 22)*

These are polymorphic, rounded, spindle shaped or pear shaped haemocytes when they are free in the haemolymph but, found flattened or elongated when attached to different organs in the insect. Their number is quite large in Lepidopterus insects. These cells are characterized by rounded and oval nucleus, grannular eosinophillic, hyaline non vacuolated cytoplasm, stained faintly bluish.

3. *Grannular cells (Fig. 22)*

These cells contain numerous reddish granules. However, their shape and structure is more or less similar to plasmocytes.

4. *Spherule cells*

The size of nucleus vary with the cells. Spherules are rarely present, particularly in the blood of cockroach. But, common in many lepidopterous insects. Cytoplasm resolution is poor.

5. *Plasmocytes (Fig. 22)*

These are bigger sized cells with oval or ovoid shape and distinct cell membrane. These are most common and most active cells found in the haemoymph. These cells are further characterized by having prominent rounded nucleus deeply stained, grannular eosinophilic and 9.12 μ in diameter. Cytoplasm is a huge bulk of basophilic granules.

Results

Following types of haemocytes were present in the blood of cockroach:

1. Plasmocytes.
2. Spherules.
3. Spindle shaped plasmocytes.
4. Proleucocytes.
5. Granular cells.

In *H. armigera* oenocytes were present as additional to cockroach.

10.

Quantitative Count of Haemocytes from Blood of a Given Insect

Experimental Insect

Cockroach *Periplanata americana* (Blattaria : Blattidae)

Requirement

Apparatus

Neubauer's Haemocytometer (1.5 WBC) (Fig - 23), counting slide (Fig - 24 & 25) with special coverslip, pipette, cotton needle, hot water bath, beaker 500 ml, thermometer, microscope.

Chemicals

Diluting fluid, (Versene), distilled water.

N.B: Haemocyte number vary with the different stages of the insects and with the differnt species.

Procedure

Take the cockroach and dip it into hot waterbath (temperature 55 to 60°) for about 1 to 2 minutes and then take a cut to the terminal point of antenna. Then collect the blood from antenna as a drop by sucking it through a clear pipette. The blood is sucked upto the mark one and immediately it is diluted in physiological saline – versene upto 101 ml. Then shake the solution for about 20 minutes so that the haemocytes will be distributed uniformly. Then take clear counting slide and a drop of diluted haemolymph solution and immediately covered with the special type of coverslip. Now observe the slide under microscope. The picture will show that haemocytes are uniformly dispersed on squares of the counting slide. Now count the number of cells present in each big square and calculate the average, volume of each square and then 1 mm^3 is also calculated.

Calculations

Formula: No. of haemocytes per mm^3 = 25,000 × n

n = is average no. of cells present in 1 sq^3.

Volume of square = length × breadth × depth

$$= \frac{1}{5} \times \frac{1}{5} \times \frac{1}{10} = \frac{1}{250}$$

Thus each square having volume $\frac{1}{250}$ mm^3 when the average no. of cells present in each square is = n.

Then ∵ 100 times dilution contains 250n cells/cubic mm

∵ if no dilution then 250 × 100

= 25,000 n

Hence, the formula is = 25,000 n

When n is the number of cells present in one square and dilution is 100 times.

Therefore, average number of haemocytes present in one sq is

$$= \frac{\text{Total no. of haemocytes present in all counting square}}{\text{No. of all sq}^3 \text{ counted}}$$

$$n = \frac{36}{16} = 2.25$$

$$= 2.25 \times 25{,}000 = 56{,}250/mm^3$$

Results

The number of haemocytes present in one cubic mm area is *56,250.*

Thus, 56,250/mm^2/haemocytes are present in a given insect haemolymph.

Precaution

1. The count should be made immediately.
2. Dilute solution should be kept in convenient container and

place for immediate use.

3. Greatest care should be taken for preparing dilutions.
4. The chamber of coverslip should be dry and clear.
5. The counting chamber should be washed with distilled water and dried with soft fabrication.

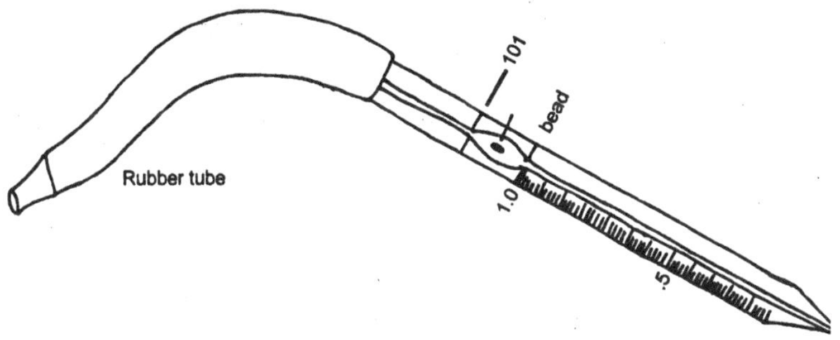

Fig. 23 : Haemocytometer

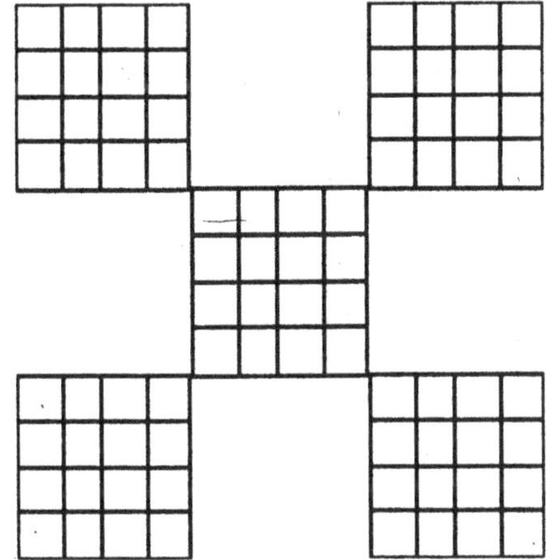

Fig. 24 : Haemocyte counting slide

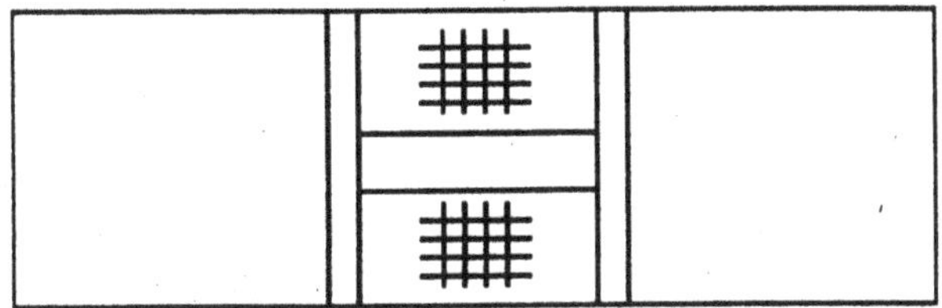

Fig. 25 : Haemocyte counting slide

Observations

No of corner	No of cell sq counted	No of leucocytes present in each sq	Total squares counted = 16
Ist	1	1	
	2	2	No of leucocytes present in sq = 8
	3	3	
	4	2	
IInd	1		
	2	2	No of cells present in sq = 9
	3	3	
	4	2	
IIIrd	1	3	
	2	2	Total cells present = 10 in sq
	3	3	
	4	2	
IVth	1	3	
	2	2	Total cells present = 9 in sq.
	3	3	
	4	1	

Total no of sq counted = x = 16

Total no. of cells counted = y = 36 ($\overset{I^{st}}{8} + \overset{II^{nd}}{9} + \overset{III^{rd}}{10} + \overset{IV^{th}}{9} = 36$)

Average no of cells present in each sq = $y/x = \frac{36}{16} = 2.25$

Hence the total no of cells = 2.25 × 25,000 = 56,250/mm^3

11.
Chromatographic Separation of Amino Acids from Haemolymph of a Given Insect

Experimental Insect

Cockroach *Periplanata americana.* (Blattaria : Blattidae)

Requirements

Apparatus

Circular chromatographic paper (Fig - 26) micropipette, petridish pair, beaker 500 ml, table lamp etc.

Chemicals

Standared amino acids, solvent system, Ninhydrin, N-butanol = acetic acid = distilled water (4 : 1 : 1).

Importance and Theory

Twelt (1908), first introduced the chromatography. In paper chromatography. Paper acts as carrier of aqueous phase of the two phases solvent system. The chromatography depends on method of absorption, partition and ion exchange. By absorption method, different components of solutes are separated. Partition part is carried out on paper and ion exchange depends on exchange of either cations or anions in chromatography method. The solute is separated into two different components moving at various rate along the paper depending upon alternate partition coefficient, solvent sytem, direction of flow and tvpe of paper used. The rate ~f flow is ratio of linear movement of solute to linear rate of movement of solvent system.

$$\text{Rate of flow (R.F.)} = \frac{\text{Distance travelled by solute}}{\text{Distance travelled by solvent}}$$

Amino acids are identified with the help of paper chromatography using solvent system containing nynhydrin, N - butanol, acetic acid and distilled water in the ratio 4 : 1 : 1. With the help of R.F. values, the amino acids can be identified as after separation.

Procedure

Take a circular chromatographic paper and draw circle at centre. Then mark various points for loading known amino acids. Likely, a unknown points also marked for loading haemolymph of nymph and adult cockroach.

Loading

At different points now load the various aminoacids and unknown samples of haemolymph. During loading drying is also essential. Drying is done with the help of table lamp. Minimum 10 drops should be loaded at each known point and simultaneous drying is done on table lamp and finally unknown points are loaded and dried as suggested above.

Spreading

After loading and drying make a hole at centre through which a paper strip is inserted and dipped into solvent in petridish which is in sufficient amount. Later, make it air tight by applying grease along the end of the chamber.

Developing

After sufficient running take out paper mark solvent front and dry it over lamp. Then pour ninhydrin 0.1% on that paper and again dry it. Coloured bands of amino acids will be developed on paper. Mark it and calculate the Rate of Flow (R.F) and finally compare with standard aminoacids and confirm the aminoacids present in unknown samples. The above procedure is repeated for compairing larva, pupa and adult of the insect.

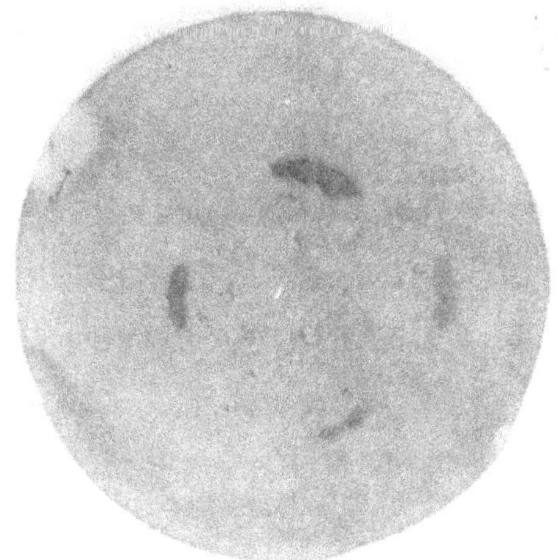

Fig. 26 : Circular chromatographic paper with bands

Observations*

Spot	Standard Amino acid	Distance travelled by solute	Distance travelled by solvent	R.F. Value
A	Cystine	0.5	5.0	0.10
B	Valine	3.0	5.0	0.60
C	Serine	1.5	5.0	0.30
D	Leucine	3.7	5.0	0.74
E	Alanine	2.0	5.0	0.40
F	Histidine	0.33	5.0	0.06
G	Unknown	0.5	5.0	0.10
	Larva	1.5	5.0	0.30
		2.0	5.0	0.40
		3.7	5.0	0.74
H	Unknown	0.5	5.0	0.10
	Adult	1.5	5.0	0.30
		3.0	5.0	0.60
		3.7	5.0	0.74

* Hypothetical

12.

Demonstration of Neurosecretory Cells in a Brain of a Given Insect

Experimental Insect

Cockroach *Periplanata americana* (Blattaria : Blattidae) or Blister beetle, *Zonabria pustulata* (Meloidae : Coleoptera)

Requirements

Apparatus

Microscope, dissection box, dissecting tray, table lamp, slides and coverslips, etc.

Chemicals

Ringer's solution, Bowins fluid, alcohol grades, 2.2% $KMNO_4$ 2.2% H_2SO_4, sodium metasulphide, methyl benzoite, paraldehyde fuschin, canada balsum, etc.

Importance and Theory

Neurosecretory cells are present in the brain (Fig. 27), corpora cardiaca, corpora allata, individual ganglion in cockroach and even on foregut in certain insects like locust. The neurosecretory cells are responsible for either producing hormones or act on other endocrine organs to stimulate them to produce hormones, thus, acting as intermediate between the nervous system and endocrine gland.

There are two groups of neurosecretory cells present on each side of the brain. One group is a pair of intracerebralis near the midline and axon. From these group nerve passes backwardly through the brain and turn back to corpus cardiacum and to corpus

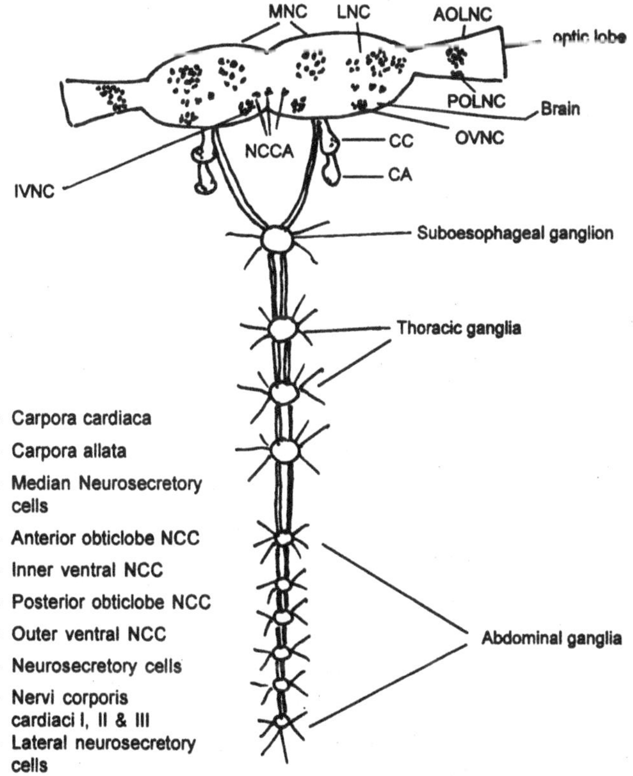

CC – Carpora cardiaca
CA – Carpora allata
MNC – Median Neurosecretory cells
AOLNC – Anterior obticlobe NCC
IVNC – Inner ventral NCC
POLNC – Posterior obticlobe NCC
OVNC – Outer ventral NCC
NCC – Neurosecretory cells
NCCA – Nervi corporis cardiaci I, II & III
LNC – Lateral neurosecretory cells

Fig. 27 : Neurosecretory cells and nerve system of cockroach

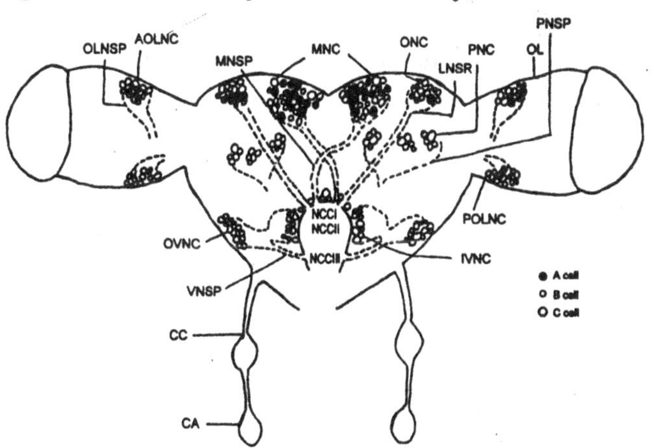

Fig. 28 : Brain and neurosecretory cells

allatum. Second group of neurosecretory cells is located at different position, it may be medial to the pendunculata or between the lateral or optic lobe. In praying mantids and other phasmids a third group of cells is situated on the deuto and tritocerebrum. In cockroach there is also a group of neurosecretory cells located on the suboesophageal ganglion which is involved in secreting hormone for rythmic locomotary activity. It is known that the neurosecretory cells increases in number during sexual maturity and for aging activity in honey bees.

Neurosecretory cells are mostly large sized and lobulated. They contain granular cytoplasm or inclusions that take deep stains of paraldehyde fuschin or chromohaematoxylin after permanganate oxidation.

Procedure

Dissect out the brain of the given insect in Ringer's solution. Then fix it in Bowins fluid for 24 hours. Later, give the wash of 70% alcohol and then dehydrate through various grades of alcohol (30 to 70%) for 5 minutes. Then keep the tissue (Brain) in 2.2% $KMNO_4$ and then in 2.2% H_2SO_4 for few minutes. Then wash the tissue with distilled water (two changes). Again dehydrate through alcoholic grades, 30%, 50% & 90% for 5 minutes each. Then stain the tissue with paraldehyde fuschin for 3 minutes. Extra stain may be washed with 90% alcohol. The tissue is then passed through 2 to 3 changes of absolute alcohol for 5 minutes and later, treated with methyl benzoite and then mounted in canada balsm on the slide for observation under microscope.

Observations

Three groups of neurosecretory cells are located on the brain of cockroach which are shown in the Fig. 27 and 28.

Results

Neurosecretory cells locted on the brain at following positions:

1. MNC
2. LNC
3. AOLNC + POLNC
4. OVNC
5. Others

13.

Detection of Enzymes from Salivary Gland of a Given Insect

Experimental Insect

Cockroach *Periplanata americana* (Blattaria : Blattidae)

Flour mill beetle – *Platynotus belli* (Coleoptera : Tenebrionidae)

Grasshopper – *Hieroglyphus banian* (Orthoptera : Acrididae)

Requirements

Apparatus

Dissection box, watch glasses, filter paper, mortar and piston, test tubes, beaker, waterbath, etc.

Chemicals

2% Starch, iodine, ethanol, Kcl, Nacl sol^n, $Cacl_2$, $NaHCO_3$, NaH_2PO_4, distilled water etc.

Importance and Theory

Enzymes play an important role in various kinds of metabolic processes as biocatalyst. The enzymes acts on specific substrate thus, they are specific in their action. Enzymes are present in salivary gland, gut and malpighian tubes in insects. Digestive enzymes are related to diet and species of the insects. Cockroach and grasshopper secretes Amylase, Protease, Lipase, Invertase and Maltase in gut as they are omnivorous while, Protease and Amylase are dominent in certain orthopterous predaceous insects (*e.g. Decticus* sp). The enzyme amylase is absent in some races of silkworm *Bombyx mori* and in honey bees. Detection of qualitative and quantitative activity of an enzyme is possible and very easy.

Procedure

Dissect out salivary gland and reservior and rinsed them in cockroach ringer solution. Then the matter is grind up in 0.5ml distilled water. Then filter the solution and the filtered extract is taken and tested for enzyme activity.

Tests for Enzymes

Amylase

Amylase test includes following steps:

1. Incubation of extract with 2% starch solution at 40°C for 5 min.
2. Testing for degradation of starch solution at intervals with acidified iodine solution.
3. Colour of starch iodine mixture will change from blue to clavet. This is indication of convertion of it into dextrins and matax by the action of amylase and hence the presence of amylase in the extract.

Maltase

Incubation of extract upto loss of colour will indicate degradation of iodine to glucose due to the action of maltose. This generation of glucose is tested by the preparation of osazone of the prolonged incubation. Glucosazone is insoluble but maltosazone is very soluble in hot water.

Protease

Incubate the salivary and gut extract for many hours at 40°C with calcified milk. The calcified milk is made acidic by adding 0.01 N HCl to it. PPT will be developed. This will indicate the presence of protease.

Lipase

Incubate the extract for 30 minutes at 40°C with olive oil dissolved in ethanol. The extract is then made alkaline by adding 2% sodium carbonate. This will yield acidic product, fatty acid, and this will confirm the presence of lipase.

Lactase

Incubate the extract for 30 minutes at 56°C with 1% lactose solution. Glucose is confirmed by the osazone test. Glucosazone is not soluble in hot water while, Lactosazone is very soluble in hot water.

Invertase

Incubate the extract at 50°C for 5 minutes with 1% sucrose solution. The reducing sugars, glucose and fructose will be formed which were confirmed by fehlings test.

Fehling's Test

Take 2ml fehling's solution A and B each in a test tube, mix and boil and add few drops of incubated extract. Production of yellow or brownish red PPT indicates the presence of invertase.

$$\text{Sucrose} \rightarrow \text{glucose} + \text{fructose}$$

Results

In the extract of salivary gland and reservior following enzymes have been detected:

1. Amylase, maltase and invertase (detected readily)
2. Protease detected in small amount.
3. Lactase and Lipase could not be found in extract of salivary gland of cockroach.

14.
Detection of Digestive Enzymes from the Gut of a Given Insect

Experimental Insect

Grasshopper *Hieroglyphus banian* (Orthoptera : Acrididae) or Cockroach *Periplanata americana* (Blattaria : Blattidae)

Requirement

Apparatus

Dissection box, test tubes, centrifuge machine etc.

Chemicals

10% NaOH soln, 0.65% NaCl soln, 0.5% $CuSO_4$ soln, 0.01% Iodine soln, 0.1% starch soln, etc.

Procedure

1. Extract of gut should be prepared with freshly collected insects.
2. Dissect out gut of given insect under icecold water.
3. Remove alimentary canal from adjoining tissues and fat bodies.
4. Wash the alimentary canal with distilled water.
5. Locate the three parts, *i.e.* foregut, midgut and hindgut of alimentary canal.
6. These parts are grinded separately in the saline solution.
7. Centrifuge the grinded parts separately for about 10 minutes.

8. Obtain clean supernant liquid and debris of tissue in test tube after centrifugation.
9. The clear liquid is taken out in test tube and stored in the refrigerator.
10. Use the extracts of gut for testing presence of enzymes. Before testing, substrate and substance are incubated at 37°C for 10 minutes. In control reading the extract is boiled for half an hour.

Tests for enzymes

1. Protease and Peptones

Take 2–3 ml of digestive extract in test tube, add equal amount of 10% NaOH to it. Mix the solutions evenly and add 0.5% Cu_2SO_4 drop by drop till purplish violet colour produced. This will confirm presence of protease and peptones in the extract.

2. Sucrase

Take equal amount (5 ml) of 10% NaOH and sucrose solution in a test tube. 1 ml of digestive extract is added to this mixture and then the mixture is kept for 2 hrs and boiled for denaturation of protein and then filtered. Later, filtrate is tested with Benedicts solution. Benedicts solution reduced while in control, reduction is not observed.

$$\text{Sucrose} \xrightarrow{\text{sucrose enzyme}} \text{Glucose + Fructose}$$

(Reduction to White)

3. Amylase

Take 1 ml of gut extract (homogenous) add equal amount of 3% starch solution to the extact. Then add 1 or 2 drops of toluene and the mixture is incubated for 1 hour at 37°C. The reaction is stopped by adding 1% TGA. Nelson and Somogys methods for sucrose estimation are adopted taking 1 ml of supernant.

4. Lipase

As a substrate, an emulsion of olive oil is taken and mixed with 1 ml of 0.1% $CaCl_2$ and further few drops of 1% NaOH are added to 2 ml of extract. As a indicator phenopthalene is added. Pink coloured solution is obtained which is then incubated for 48 hours. The fraction of pink colour solution to transparent colour indicate transformation of fats to fatty acids. For lipase test rectum from the gut is totally excluded.

5. *Starch*

Take 2 ml of extract and 5 ml of starch, paste in test tube and incubate. The solution is divided into two parts.

I[st] part + Iodine solution will provide deep blue colour which will confirm the undigested starch.

IInd part + Benedicts soln→soln reduced confirmed by PPT. The amount of PPT is related to presence of sugar in solution.

Amylase react with starch.

Reaction

Starch $\xrightarrow{\text{Amylase}}$ Maltose sugar $\xrightarrow{\text{Maltase}}$ Reducing sugar $\longrightarrow$ Reduces Benedicts solution

Observations

Sr No.	Name of enzyme	Test	Fore gut	Mid gut	Hind gut
1	Protease	Burate	–	+	–
2	Sucrase	Benedicts	–	–	+
3	Lipase	Olive oil	–	–	+
4	Maltase	Acidified Iodine	+	+	+

+ = Present

– = Absent

Results

Sr No.	Enzymes present in gut part	Enzyme	Amount
1	*Fore gut*	Lipase Maltase	Higher Normal
2	*Mid gut*	Protease Maltase	Small Higher
3	*Hind gut*	Sucrase Maltase Lipase	Small Small Higher

15.
TO ESTIMATE ENZYMES FROM A GIVEN INSECT

Experimental Insect

Flour mill beetle, *Platynotus belli* (Coleoptera : Tenebrionidae)

Requirements

Apparatus

Dissecting box, beakers, test tubes, pipettes, measuring cylinders, weight box, chemical balance, chillded mortar and piston, hot water bath, pH meter, spectrophotometer, thermometer 400°C etc.

Chemicals

Reagents

A. Substrates

1. For amylase 1% starch (soluble) in distilled water.
2. For invertase 2% sucrose in distilled water.
3. For trehalase 1% trehalase in distilled water.

B. Buffer

1. Phosphate buffer 7.2 pH

C. DNSA Reagent

19 gram of 3, 5 dinitrosalicylic acid is dissolved in 20 ml of 2 N NaOH, add to this 30 gms of potassium sodium tartarate,

shake well and obtain clear solution. Then add distilled water, strength = 1mg/ml.

Importance and Theory

Many enzymes take part in digestion process of an insect species. The enzymes are related to types of the food consumed, pH, temperature, substrate concentrations, time, activators, inhibitors, etc. The food of the present insect species largely comprises of carbohydrates (flour of cereals). Hence considered for the present study.

Amylase, invertase and trehalase, starch and sucrose are major components in the cereal flour. Trehalose, the major blood storage sugar in insect, which is synthesized by the insects from monomeric units, glucose present in the blood. The midgut is the main site of enzymatic secretion and digestion in insects.

Amylase, invertase and trehalase activities are determined by using 3, 5 dinitrosalicylic acid reagent (DNSA) for determining the reducing sugars, *i.e.* the free aldehydic groups of glucose formed after starch or sucrose or trehalose digestion. This reaction is based on the reduction of dinitrosalicylic acid by the aldehydic group of glucose units in basic medium. The reduced DNSA is a colour complex and its colour is measured spectrophotometrically at an absorbance of 540 nm.

Procedure

1. *Tissue preparation*

Dissect out beetle in chilled insect ringer and take out midgut, then weigh the midgut tissue carefully and homogenise it in chilled 0.75% NaCl by using chilled glassmortar and piston. Centrifuge the content for 15 minutes at 5000 × g and measure the volume of supernant. Use aliquates of supernants as enzyme sourse, which is used as the concentration of 1 gm/ml.

(1 gm/upto 20 ml is also allowed)

2. *Assays*

Assays for amylase, invertase and trehalase are arranged separately in triplicate with their respective controls. Reaction mixture containing 1 ml of phosphate buffer (pH. 7.2), 1 ml of their respective substrates and 0.5 ml of enzyme as extract is taken in test tube. In control, enzyme extracts are added after termination of reaction. Incubate the reaction mixture in waterbath at 350°C. Incubation period for amylase, invertase and trehalase is 15 minutes, 60 minutes and 60 minutes respectively. During incubation, test tubes should be shaked frequently. After incubation, terminate the enzymatic reactions by adding 2.5 ml of DNSA reagent. Again the tubes are heated in waterbath at 100°C (boiling) for 5 minutes, followed by immediate cooling (by tap water). Then add 2.5 ml distilled water and mix the content. Finally, determine the enzyme activity with the help of spectrophotometer by measuring the optical densities at 540 nm of the colour developed in reaction mixture against their respective blanks. OD obtained are compared with standard graph for confirming amylase is equivalent to maltose while invertase and trehalase are equivalents to glucose.

3. *Preparation of Standard Curves*

Standard curves are obtained from the direct reaction of maltose (0.1 mg to 2 mg) in reaction mixture) for amylase and glucose (0.1 mg to 2 mg in reaction mixture) for invertase and trehalase with DNSA reagent under the condition similar to those for enzyme reactions. The activities are expressed mg maltose/mg tissue/hour for amylase and mg glucose/mg tissue/hour for invertase and trehalase.

Standard graphs

Amylase							Invertase and Trehalase						
Tube No.	Maltose (ml)	Maltose conc. (mg)	Dist. Water (ml)	DNSA (ml)	Dist Water (ml)	O.D.	Tube No.	Glucose (ml)	Glucose conc. (mg)	Dist Water (ml)	DN SA	Dist. water (ml)	O.D.
1.	Nil	Nil	2.5ml	2.5ml		0.0	1.	Nil	Nil	2.5ml	2.5ml	2.5ml	0.0
C	O	N	T	R	O	L	C	O	N	T	R	O	L
2	0.2	0.2	2.3	2.5	2.5		2	0.2	0.2	2.3	2.5	2.5	
3	0.4	0.4	2.1	2.5	2.5		3	0.4	0.4	2.1	2.5	2.5	
4	0.6	0.6	1.9	2.5	2.5		4	0.6	0.6	1.9	2.5	2.5	
5	0.8	0.8	1.7	2.5	2.5		5	0.8	0.8	1.7	2.5	2.5	
6	1	1	1.5	2.5	2.5		6	1	1	1.5	2.5	2.5	
7	1.2	1.2	1.3	2.5	2.5		7	1.2	1.2	1.3	2.5	2.5	
8	1.4	1.4	1.1	2.5	2.5		8	1.4	1.4	1.1	2.5	2.5	
9	1.6	1.6	0.9	2.5	2.5		9	1.6	1.6	0.9	2.5	2.5	
10	1.8	1.8	0.7	2.5	2.5		10	1.8	1.8	0.7	2.5	2.5	
11	2	2	0.5	2.5	2.5		11	2	2	0.5	2.5	2.5	

0.5 OD = 500 μg maltose → Amylase

0.5 OD = 330 μg glucose → Invertase

Calculations

1. *Amylase* :

0.5 OD = 500 μg maltose by standard graph.

we have 0.75 OD.

if 500 μg maltose is for 0.5 OD

then how much maltose for 0.75 OD?

$$= \frac{0.75 \times 500}{0.5} = 750 \text{ μg maltose.}$$

This maltose is for 10 minutes then how much for 60 minutes

$$= \frac{60 \times 750}{10} = 4500 \text{ μg maltose}$$

Wt. of tissue in 220 mg in 25 ml then how much tissue in 1 ml.

$$\frac{1 \times 220}{25} = 8.8 \text{ mg/ml.}$$

If 8.8 mg liberates 4500 µg maltose/M/homogenate then how much for 100 mg.

$$\frac{1000 \times 4500}{8.8} = 511364 \text{ µg maltose i.e. } 511364 \text{ mg/gm/hr}$$

Result = 511364 mg/gm/hr

2. Invertage :

0.5 OD = 330 µg glucose

Hence for 0.12 OD how much

$$\frac{0.12 \times 330}{0.5} = 792 \text{ µg glucose/30 min.}$$

∴ 792 µg glucose for 30 min.

∴ for 60 min. = 1584

∴ 1584.00 µg glucose for 8.8 mg tissue

$$\therefore \text{ for } 1000 \text{ mg} = \frac{1000 \times 1584}{8.8} = 1{,}80{,}000 \text{ µg glucose}$$

Result

180 mg glucose/gm/tissue/hr.

16.
TO FIND OUT LETHAL DOSE (LC 50) OF INSECTICIDE IN INSECT

Experimental Insect

Cockroach *Periplanata americana* (Blattaria : Blattidae)

Requirements

Aparatus

Insect cages, spray pump, stop watch, measuring cylinder, beakers, petridishes etc.

Chemicals

Insecticide, kerosene, Distilled water, etc

Importance and Theory

Insecticides are an important component of pest management, without insecticides pest mangement is still practically incomplete. Although other control measures also play an important role they have certain limitations. However, there is need to minimise the use of insecticides in pest control programme because, insecticides cause several serious problems like pollution and health hazards, pest resistance, pest resurgence, secondary pest out break, etc. Appropriate concentration and characteristics of insecticide, environment and type of ecosystem, insecticide appliances, etc. counts the efficacy of insecticide. Thus, treating the crops with appropriate concentration will help in reducing the use of pesticides and in keeping the environment ecofriendly upto certain extent.

Procedure

Take a given number of insects 30 in different insect cages (Fig. 10 & 11). Dilute the given insecticides with either solvent or water (as per the characteristics of pesticides). Initially make four concentrations, 25%, 50%, 75% and 100% and treat the insects in cages. Observe the mortalities in each concentration. If the mortality is more than 50% then go for dilute concentrations till 50% mortality is obtained and thus confirm the Lc 50. Note down the deaths of each insect after spraying the insecticide for 30 minutes. Repeat the same procedure five times for confirming the results and plot the graph against mortality and per cent concentration of the insecticide.

The per cent of mortality (Lc 50) can be calculated by using Abbort's formula :

$$\frac{v-b}{v} \times 100 = \% \text{ mortality}$$

Where,

1. V is number of insects taken for the experiment.
2. b is the number of living individual after spraying at particular time.
3. V - b is the number of individuals killed by insecticide at particular time.

Observations

Sr. No.	Time (min)	Number of indiviuals died at each concentration			
		25%	50%	75%	100%
	1	0	0	0	0
	2	0	0	2	3
	3	1	2	4	6
	4	1	3	5	11
	5	2	4	7	13
	6	3	5	10	14
	7	4	7	13	15
	8	5	9	15	17
	9	7	10	16	18
	10	8	11	18	22
	11	9	12	20	24

Sr. No.	Time (min)	Number of indiviuals died at each concentration 25%	50%	75%	100%
	12	10	13	23	28
	13	12	15	25	30
	14	13	16	26	
	15	14	18	28	
	16	15	20	29	
	17	16	22	30	
	18	18	23		
	19	19	24		
	20	20	25		
	21	21	26		
	22	22	28		
	23	23	29		
	24	24	30		
	25	25			
	26	26			
	27	27			
	28	28			
	29	29			
	30	30			

Results

Out of 4 concentrations tested 50% concentration was found as Lc 50.

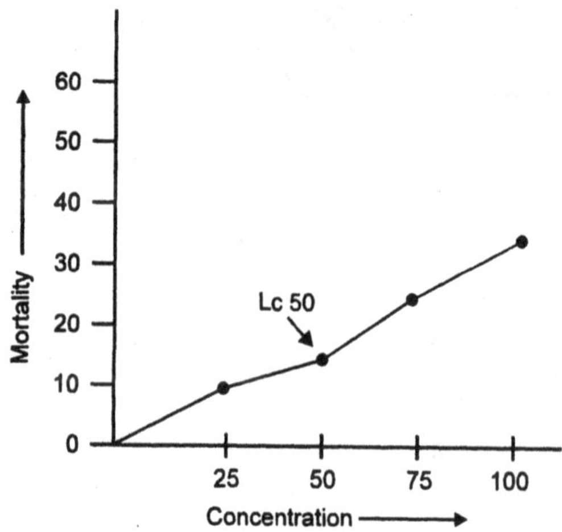

Fig. 29 : Per cent mortality in a given insect.

Calculations

1. For 25% = $\frac{V - b}{v} \times 100 = \frac{18}{30} \times 100 = 40\%$

2. For 50% = $\frac{V - b}{v} \times 100 = \frac{15}{30} \times 100 = 50\%$

3. For 75% = $\frac{V - b}{v} \times 100 = \frac{25}{30} \times 100 = 83.33\%$

4. For 100% = $\frac{V - b}{v} \times 100 = \frac{30 \times 100}{30} = 100\%$

17.

Demonstration of Chemoreception in House Fly

(Response to Sweet–Sucrose/Glucose)

Experimental Insects

House fly *Musca domestica* (Diptera : Muscidae)

Requirements

Apparatus

Blocks of bee wax, forcep, specimen tubes, glass rod, watchglass or cavityblocks, microscope, small beakers, etc.

Chemicals

Glucose, sucrose, distilled water.

Procedure

A dozen or more living house flies are required for this experiment. Paralize the houseflies with CO_2 and fix them to small blocks of beewax (Fig. 30). An angular (90°) rod is used for fixing the house fly. The fly is fixed to the end of glass rod (Fig. 30). The flies are starved for 24 hrs between mounting and start of experiment. However, flies can survive for days if given water. 1, 0.5, 0.1, 0.05, 0.01 molar solutions of sucrose/glucose are prepared for testing flies in following ways:

2 gm glucose/sucrose in 20 ml of distilled water → 10% glucose/sucrose

1% ⟶ 1 ml ⟶ 10 ml

0.5% ⟶ 1 ml ⟶ 20 ml

0.25% ⟶ 1 ml ⟶ 50 ml

0.1% ⟶ 1 ml ⟶ 100 ml

0.05% ⟶ 1 ml ⟶ 200 ml

0.01% ⟶ 1 ml ⟶ 1000 ml

From above prepared solution take 1 ml each time.

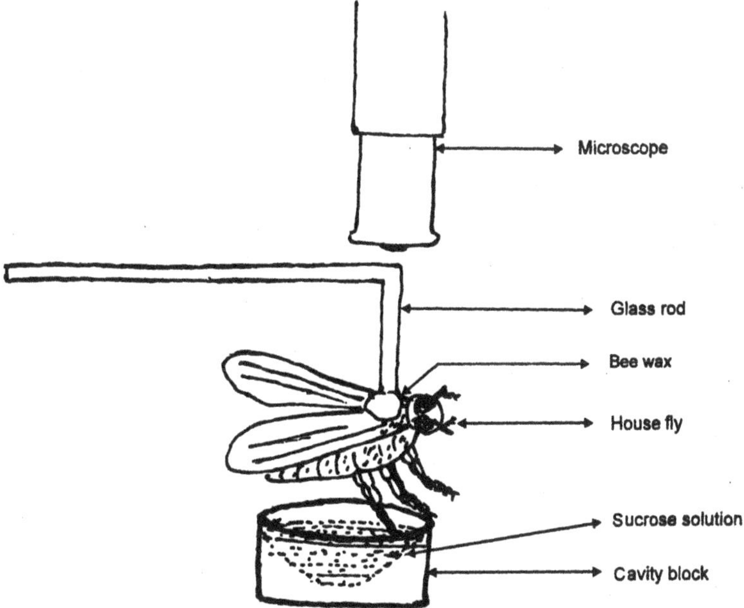

Fig. 30 : Chemoreception in house fly

Before the experiment, let the flies have water but not food for 24 hrs. Take each fly and expose to the water. After several negative responses of fly to distilled water, start testing sweet concentrations from the lowest to highest. Do not allow the fly to feed after proboscis extension. Note the threshold concentration of sugar which can be detected by the tarsal chemoreceptor. Record the threshold concentration of sucrose/glucose to each fly.

Observations

Species No	1	0.5	.1	.05	.01
	← response by flies →				
1	+	+	+	+	–
2	+	+	+	–	–
3	+	–	–	–	–
4	+	+	+	+	+
5	+	+	+	+	+
6	+	+	+	+	+
7	+	+	+	–	–
8	+	+	–	–	–
9	+	+	+	+	+
10	+	+	+	+	+
11	+	+	+	+	+
12	+	+	+	+	+

+ = Positive response

– = Negative response

Result

1% solution is most receptive for the flies.

18.
DISSECTION OF BLISTER BEETLE

TAXONOMIC POSITION OF EXPERIMENTAL INSECT

Class	Insecta
Order	Coleoptera
Family	Meloideae
Genus	*Zonabria*
Species	*pustulata*

Characters

Blister beetle (Fig. 31) is cosmopolitan in distribution. Its head is prognathus. Pro-thorax is narrower than head and both joined by neck. Compound eyes are large, and wel developed. Ocelli absent. Antennae long, simple, 11 segmented. Blister beetle measures about 1 inch in body length and 1 cm in breadth. It is black in body colouration with orange or yellow zig-zag strips on the elytra. Wings and body are softer than other beetles. Forewing is thick and called elyta. Hind wing is membranous. Legs black in colour with five tarsal segments. Abdomen black, tapering towards posterior. There is hypermetamorphosis in the blister beetles therefore, their larvae are strickingly dissimilar with each other. Beetles are pests on ornamental plants, cucurbits and several agricultural graminacious crops. They are always associated with flowering bodies of the crop. Thus, blister beetles are economically very important.

Digestive System of Blister beetle (Fig. 32)

Digestive system of blister beetle is coiled tube of different width at different regions.

Fig. 31 : Blister beetle

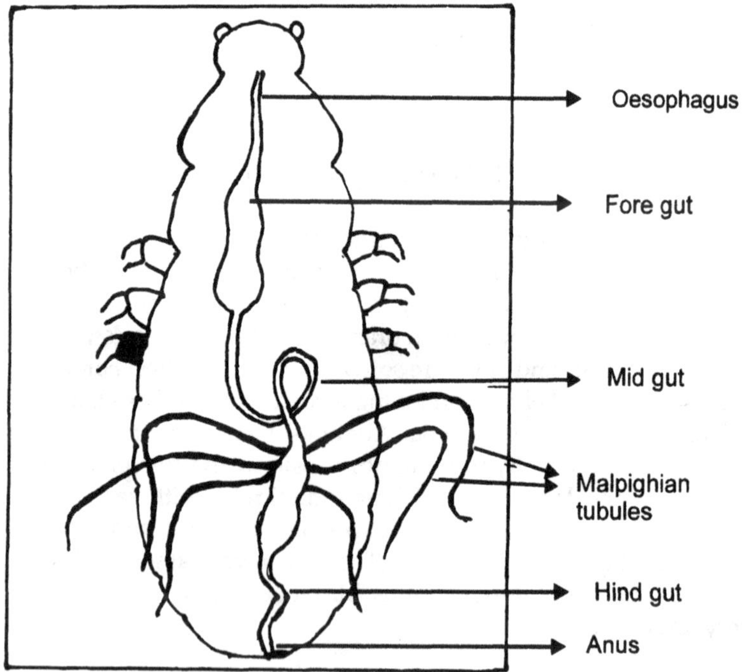

Fig. 32 : Digestive system of Blister beetle

Head is prognathus type, mouth parts lead to the oesophagus, oesophagus initialy narrow tube but further expands and that part is known as fore gut. It consequently lead to conspicuous mid-gut. Mid-gut is long and coiled tube. As hepatic ceacae are absent on the foregut, hind gut and foregut can not be firmly recognised externally. However hindgut is short. There are 4 malpighin tubules which are very long and convoluted arising from the junction of mid and hind gut. Hind gut contain rectum, colon and anus as external opening. Salivary glands and anal glands are absent.

Nervous system of Blister beetle (Fig. 33)

Nervous system of blister beetle is more generalized type. It consists of brain and ventral nerve cord.

Circum oesophageal connectives joins the brain with the ventral and to the suboesophageal ganglion. The optic and antennary nerves are given out by the brain. The 1st thoracic ganglion is big in size. There are three thoracic ganglia. IInd and IIIrd thoracic ganglia are closer and relatively smaller.

There are 4 abdominal ganglia. The last ganglion is large one and formed by fusion of many ganglia. There are several branches of nerves given out to various closely related area of the insect body.

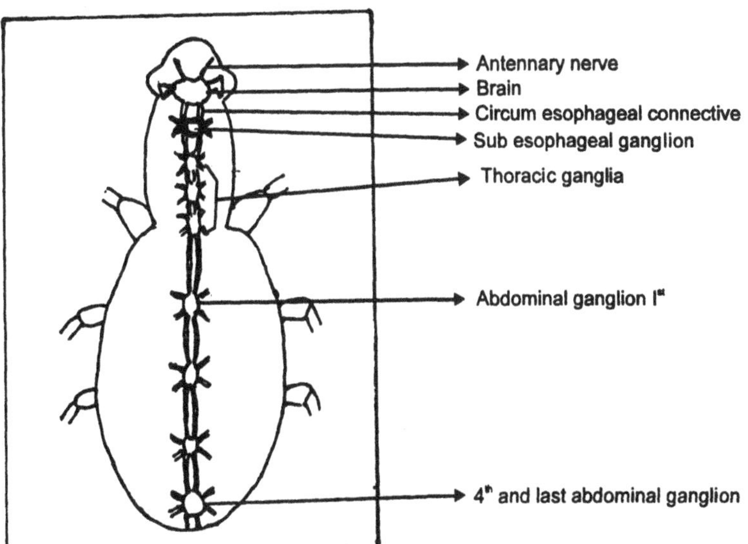

Fig. 33 : Nervous system of Blister beetle

Male reproductive system of Blister beetle (Fig. 34)

The male reproductive system consists of a pair of testis. Three pairs of accessary glands are present which open into ejaculatory duct. Testis are sessile, compound and divisible into number of follicles. All enclosed in a common sheath. Both vas deferentia open into common ejaculatory duct separately. Ejaculatory duct is long tube which enters to genital pouch. There are 3 pairs of accessary glands, 1st two pairs are at anterior end, before ejaculatory duct and one pair posterior. However, much more variations are seen in no. of accessary glands and their mode of opening.

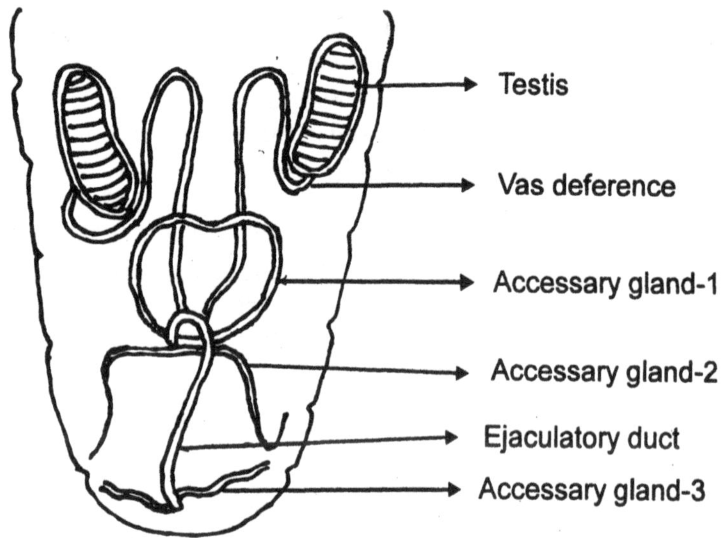

Fig. 34 : Male reproductive system

Female reproductive System of Blister beetle (Fig. 35)

Female reproductive system is composed of a pair of ovaries located in IInd abdominal segment. Each ovariole has got terminal filament. Ovarioles unite to open into oviduct. Two oviducts unite to form a common oviduct and further vagina. In vagina spermatheca opens from anterior side while, accessary gland opens from posterior side. Accessary gland is tubular in form and single. It is located on spermathica/or separately.

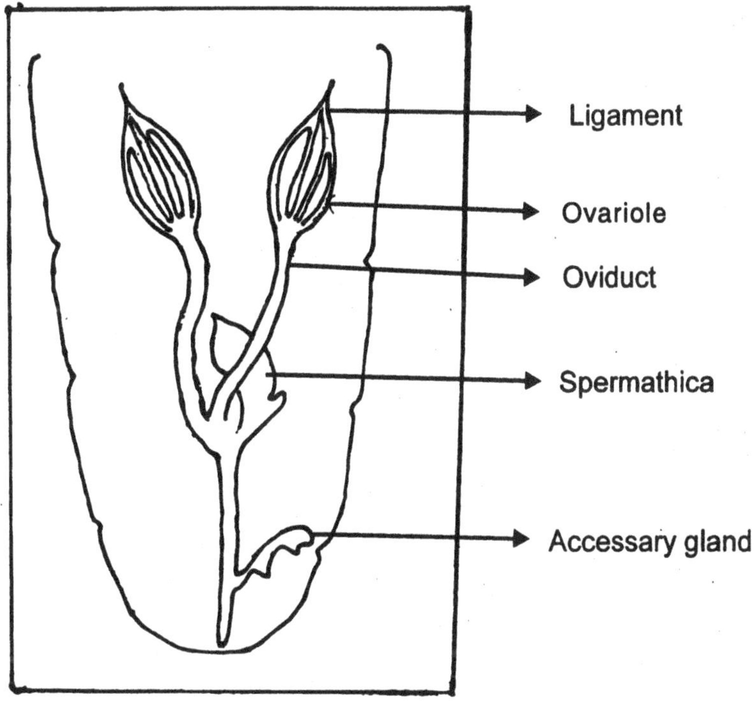

Fig. 35 : Female reproductive system

19.
DISSECTION OF PLANT BUG

TAXONOMIC POSITION OF EXPERIMENTAL INSECT

Order	- Hemiptera
Sub-order	- Heteroptera
Family	- Pentatomidae
Type	- Plant bug (Fig. 36) (Karanji kida)

Characters

Antennae long, eyes well-developed, Pulvili present, ocelli absent, meso and meta-pleurae simple. Hemielytra with corium, cuneus and clavus, nerves reticulate and membrane with 5 veins, scutellum normal, tibia without lobe, rostrum 4 segmented, tarsi 3 segmented.

Head, sclerites completely fused with each other. Labrum narrow. Mouth parts ophistho-gnathus type, proboscis is main injecting apparatus in piercing and sucking the liquid food. Pronotum is large and forms greater part of the thorax. Scutellum/mesonotum can also be seen. Male and female can be identified from the genitalia. Females are generally larger than males. Genital chamber is seen on ventral side.

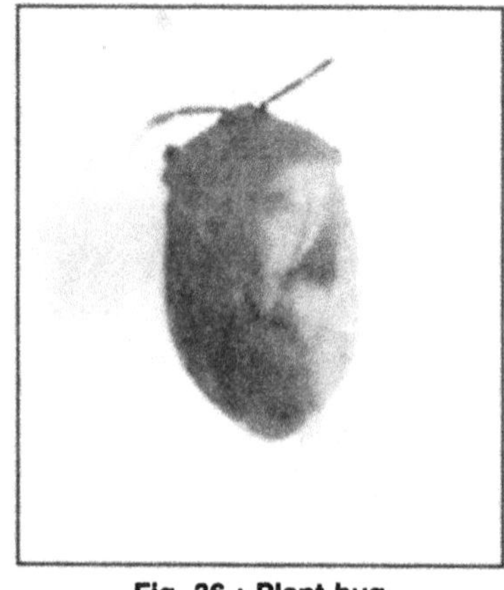

Fig. 36 : Plant bug

Digestive System (Fig. 37)

The pest feeds on juice hence the digestive canal is coiled and longer than length of the body. It consists of mouth, pharyngeal bulb, short and narrow oesophagus, long broad crop and short round gizzard. The mid gut is typical. It's I[st] part is ciliated forming a chamber. There are number of dilations. At the junction of hind gut and mid gut there are 4 malpighian tubules which opens into a dialated part of the junction. The hind gut is short. Rectum is situated in the last segment of abdomen and opens outside by anus.

Salivary glands have peculiar complex structure. They are situated in thoracic region. There are 4 lobes, two big and two small. There are 3 ducts originating from the gland which runs anteriorly as convolution.

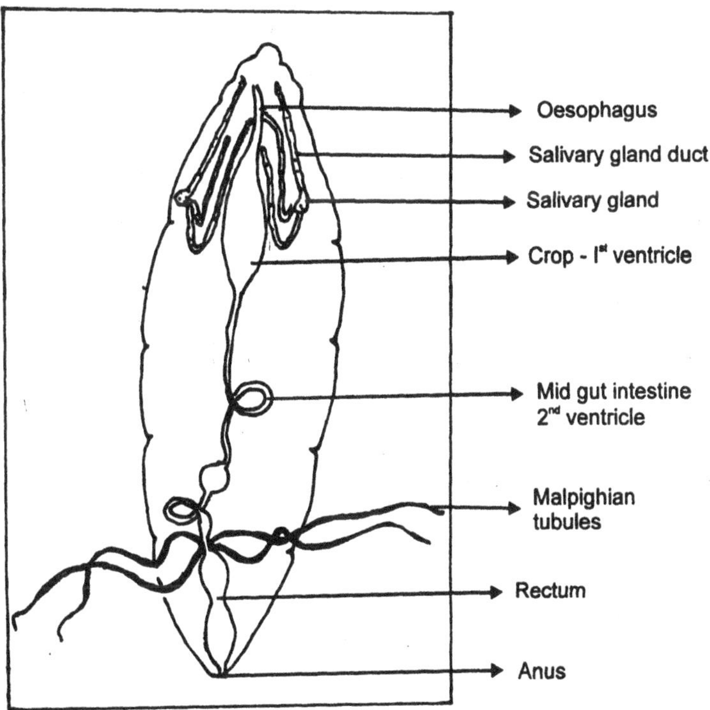

Fig. 37 : Digestive system of plant bug

Nervous system of plant bug (Fig. 38)

Nervous system of plant bug consists brain and concentrated ganglia in thorax region. The brain is located in the head region from which antennary nerves are given out. Suboesophageal ganglion is situated beneath the brain. Prothoracic ganglion is separate. Connection between prothoracic and suboesophageal, and between prothoracic and compound ganglion is double. Compound ganglion is fusion of mesothoracic, metathoracic and all the abdominal ganglia. This compound ganglion is located towards the end of the thorax (posteriorly). All the abdominal ganglia are fused with thoracic ganglia. Thus, there is reduction in ganglia in the plant bug. However, compound ganglion provides various nerves to various organs in the abdomen.

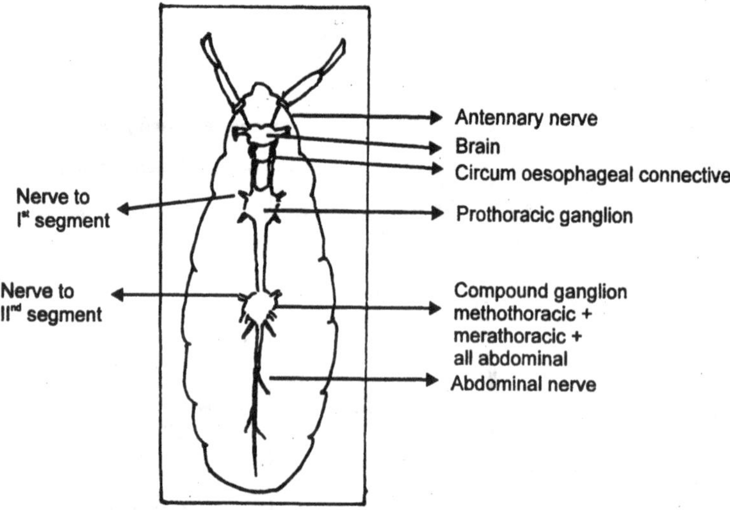

Fig. 38 : Nervous system of plant bug

Male Reproductive System (Fig. 39)

There is a pair of testis. Each testis provides a vasadeferentia which makes its course towards the posterior end, and in between 5th and 6th segment they meet and form its hollow structure on the other side to form the seminal vesicle which leads into the ejaculatory duct and finally genital chamber.

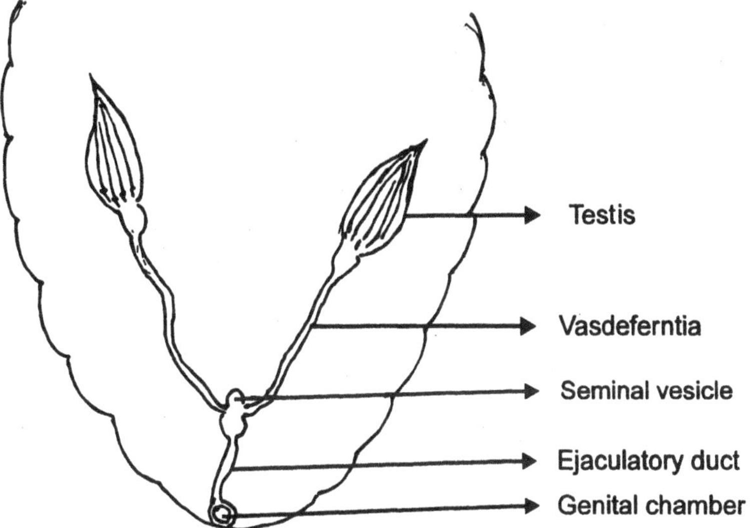

Fig. 39 : Male reproductive system of plant bug

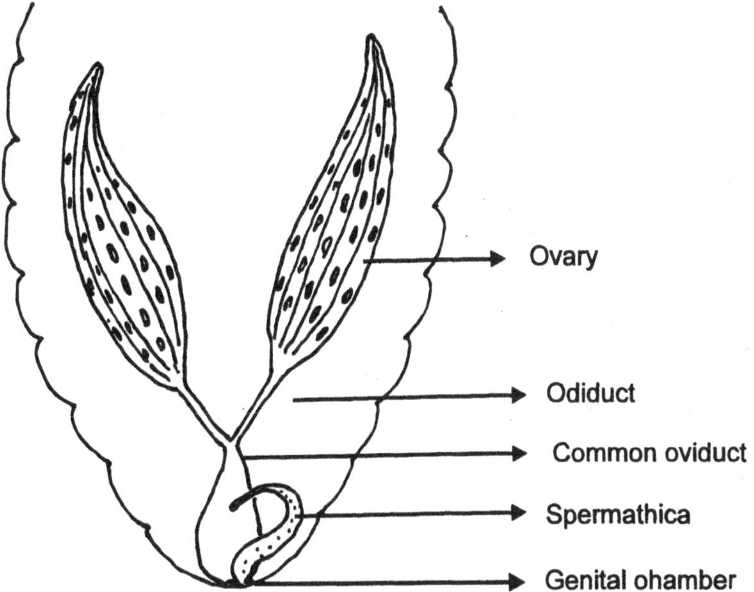

Fig. 40 : Female reproductive system of plant bug.

Female Reproductive System (Fig. 40)

There is a pair of ovary. Each ovary consists of seven ovarioles. All tapering anteriorly. These ovarioles, on one side, open into oviduct which runs posteriorly and joins with oviduct of the other side to form the common oviduct. Later, it opens into the genital chamber. Spermathica is opened in genital chamber .

20.
Dissection of Grasshopper

TAXONOMIC POSITION OF EXPERIMENTAL INSECT

Class	– Insecta
Order	– Orthoptera
Family	– Acrididae
Genus	– *Hieroglyphus*
Species	– *banian*

Characters

Adult grasshoppers (Fig. 41 & 42) show hypognathus head and filiform antennae. They measures about 40–50mm in body length and are shining greenish yellow in body colour. Wings are straight and forewing is tegmina. These are 3 black lines on the pronotum (Fig. 42) desending from dorsal to ventrolaterally. Nymphal form shows numerous reddish brown spots on their body. They are yellowish in early stages but turns greenish approaching maturity. *H. banian* is polyphagus pest of agricultural crops in India. Both, nymphs and adults cause damage to crops by feeding on their leaves. It is serious pest of rice in several states of India.

Digestive System of Grasshopper (Fig. 43)

Grasshopper is foliage feeder. Therefore, its digestive track is more or less straight tube and not coiled as like plant bug. The digestive system of grass hopper consists of hypopharynx, oesophagus, crop, gastric caecae, midgut, malpighian tubules, hind gut : colon, rectum and anus. Salivary glands are bunched and opens into oesophagus. Mapighian tubules are invariably blind

Fig. 41 : Grasshopper (Lateral view)

Fig. 42 : Grasshopper (Dorsal view)

tubes which opens into proximal part of hind gut or at the junction of midgut and hindgut.

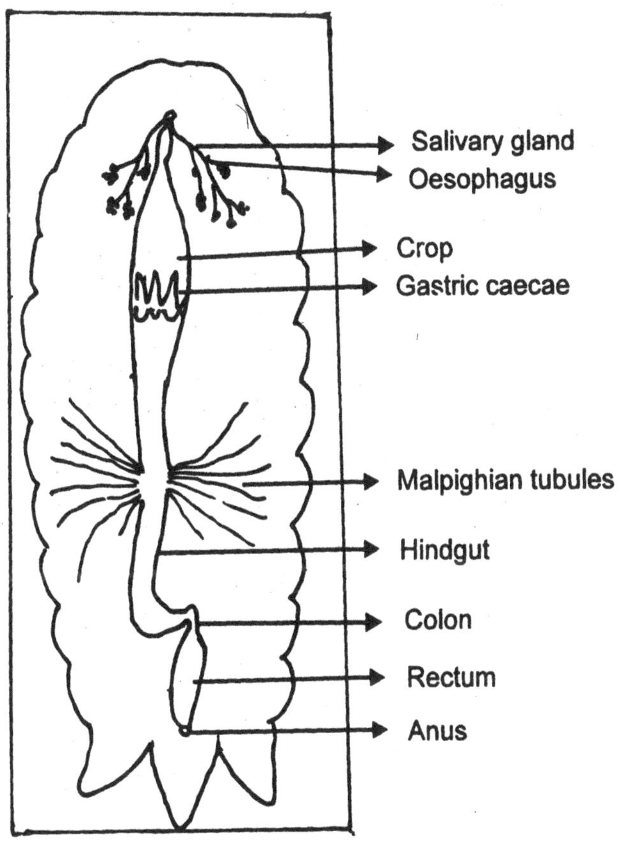

Fig. 43 : Digestive system of Grasshopper

Nervous System of Grasshopper (Fig. 44)

Nervous system of grasshopper consists Brain, subesophageal ganglion, 3 thoracic and 6 abdominal ganglia and a nerve cord.

Brain provides antennary and ocellas nerves. Oesophageal connective originates from brain and surrounds oesophagus. The ring unites to form sub-oesophageal ganglion. From sub-oesophageal ganglion nerve cord starts, which gives 3 large thoracic and 6 small abdominal ganglia. Each ganglion provides nerves to their nearer parts of the body.

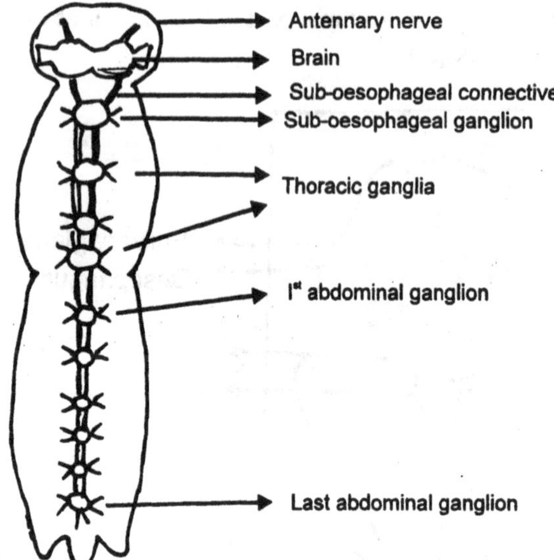

Fig. 44 : Nervous system of Grasshopper

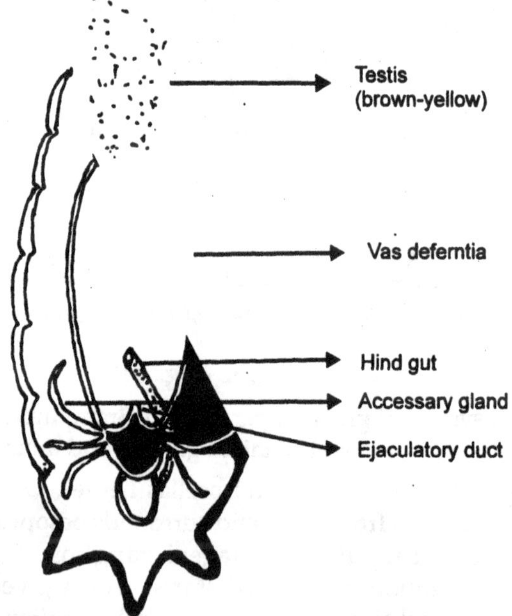

Fig. 45 : Male reproductive system of Grasshopper

Male Reproductive System of Grasshopper (Fig. 45)

Male reproductive system of grasshopper comprises testis as a combined large mass which is brownish yellow in colour, vasdeferens of a very small diameter, seminal vesicle and common ejaculatory duct. Reproducive system starts from IIIrd abdominal segment and ends at IXth segment. It opens as male genital opening. Accessary glands are present to assist and to take side role in fertilization. Circus and syncoxite are used in mating process by grasshopper.

Female Reproductive System of Grasshopper (Fig. 46)

Female reproductive system of grasshopper comprises a paired ovary with accessary glands. Ovaries are light yellow bodies which leads into corresponding oviducts. Oviducts unite to form common oviduct, further extends as genital chamber in which spermathica and accessary glands open. Posterior most part of the system is vagia which open outside. Fertilization is internal and during mating spermathica receive the sperms.

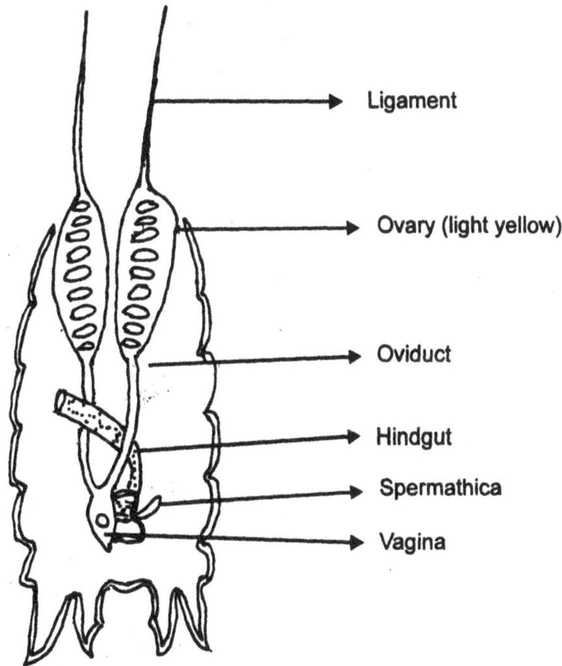

Fig. 46 : Female reproductive system of Grasshopper

Bibliography

Bergmann, M. et al. 1931. Composition of chitin. *Ber. deutsch. chem. Ges*, 64, 2436 – 40.

Berlese, A. 1900. Role of haemocytes during metamorphosis. *Zool. Anz.*, 23, 515 – 21.

Bergold, G. 1935. Form of spiracles in beetle. *Morpho. Oekol. Tiere.*, 29, 511 – 36.

Bruntz, L. 1904. Storage excretion in insects. *Arch. Biol.*, 20, 217 – 422.

Bruntz. L. 1908. Phygocytosis and excretion : Thysanura. *Arch. Zool.* 8, 471 – 88.

Buck, J. 1962. Insect respiration : Review. *Ann. Rev. Ent.*, 1, 27 – 56.

Campbell, F.L. 1929. Tests of chitin. *Ann. ent. Soc. Am.*, 22, 401 – 26.

Cuenot. L. 1895. Haemocytes, Phagocytic Organs : Orthoptera. *Arch. Zool.*, 14, 293 – 341.

Ehrenberg, R. 1914. Excretary system. *Winterstein's Handb. d.vergi. physiol.* 2, 695 – 759.

Fuller, H.B. 1960. Neurosecretary cells in ganglia of *Periplanata* and *Corethra. Zool. Jahrb. Physiol.*, 69, 223 – 50.

Gersch, M. 1942. Excretion of fluorescent substances by insects. Z. *Vergl. Physiol.* 29, 506 – 31.

Hassan, A.A.G. 1943. Spiracular mechanisms. *Trans. Roy. Ent. Soc. Lond.*, 94, 103 – 53.

Jawlowski, H.Z. 1948. Structure of insect brain. *Ann. Univ. Mariae Curie. Sklodowska* 3 (c), 1 – 37.

Johanson, A.S. 1958. Histological types of neurosecretory cells. *Nyt. Mag. Zool.*, 7, 5 – 132.

Jones, J.C. 1962. Types of haemocytes : Review. *Amer. Zoologist.,* 2, 209 – 46.

Koch, C. 1932. Tests for chitin. *Z. Morph. Oekol. Tiere;* 25, 730 – 56.

Kohnelt, W. 1955. Water relations of Tenebrionidae. *Naturuv. Kl. Abt. l* -164, 49 – 64.

Lison, L. 1937. Elimination and athrocytosis of dyes in insects. *Arch. Biol;* 48, 321 – 360.

Manunta, C. 1955. Uric acid excretion in *Bombyx mori L.* after removal of the silk glands. *Genetica ed Entomologia,* 2, 269 – 81.

Marchal, P. 1890. Uric acid excretion in insects : Hymenoptera. *Mem. Soc. Zool. Fr.* 3, 31 – 87.

de Maur, P.A. Uric acid metabolism in *Drosophila* mutants. *Z. Vererb. Lehre,* 92, 42 – 62.

Mellonby, K. 1934. Site of water loss from insects. *Proc. Roy. Soc Lond* (B), 116, 139 – 149.

Meyer, K.H. and Mark, H. 1928. Composition of chitin. *Ber. deutsch. Chem. Ges;* 61, 1936 – 39.

Nagabhushanam, R; V.R. Awad and R. Sarojini. 1981. Laboratory excercises in animal physiology. COSIP – VLP (Biolog) Publication. pp. 1 – 92.

Nishiitsutsuji - Uwo, J. 1961. Fine structure of neurosecretary cells in Lepidoptera. *Z. Zellforsch,* 54, 613 – 630.

Raabe, M.C.R. 1963. Tritocerebral neurosecretion and colour change in phasmida. *Acad. Sci.,* 257, 1171 – 73.

Sathe, T.V. and S.M. Gaikwad. 2000. Free amino acids in *Erias fabia* (Stoll) (Lepidoptea). *Geobios,* 27, 203 – 204.

Sathe, T.V. 2003. Agrochemicals and pest management. Daya Publishing House, Delhi. pp. 1 – 217.

Sathe, T.V. and Rokade, A. V. 1994. Comparative study of haemocytes of *Heliothis armigera* (Hubn.) and its parasitoiod *Campoletis chlorideae* Uchida (Hymenoptera). *Oikoassy,* 11, 23 – 24.

de Sinety, R. 1900. Malpighian tubules, & phasmidae. *Bull. Soc. Ent. Fr;* 19, 119 – 278.

Stiennon, J.A. and Drochmans, P. 1961. Fine structure of neurosecretory cells in phasmidae. *Gen. Comp. Endocrin;* 1. 286–94.

Tauber, O.E. 1934. Distribution of chitin : *Periplanata. J. Morph*, 56, 51 – 58.

Thomsen, E. 1952. Neurosecretory cells and corpus cardiacum in adult *Calliphora. J. Exp. Biol*; 29, 137 – 72.

Vinogradskaya, O.N. 1955. Role of spiracles in water loss in *Anopheles. Ent. obozr*, 33, 157 – 60.

Wago, H. 982. *Dev. Comp. Innunol*; 6, 591-599.

Wester, D.H. 1910. Tests for chitin. *Zool. Jarhb. Syst*; 28, 531 – 57.

Wigglesworth, V.B. and Gillett J.D. 1936. Loss of water at moulting: *Rhodnius*, Hemiptera *Proc. Roy. Ent. Soc.* Lond. A. 2, 104 – 107.

Index

www.ingramcontent.com/pod-product-compliance
Lightning Source LLC
Chambersburg PA
CBHW050216270326
41914CB00003BA/432
* 9 7 8 9 3 5 1 2 4 3 7 8 6 *